# SURFACTANT-BASED SEPARATION PROCESSES

# SURFACTANT SCIENCE SERIES

### CONSULTING EDITORS

*MARTIN J. SCHICK*
Consultant
New York, New York

*FREDERICK M. FOWKES*
Department of Chemistry
Lehigh University
Bethlehem, Pennsylvania

OTHER VOLUMES IN PREPARATION

# SURFACTANT-BASED
# SEPARATION PROCESSES

*edited by*

## John F. Scamehorn
## Jeffrey H. Harwell

*Institute for Applied Surfactant Research*
*University of Oklahoma*
*Norman, Oklahoma*

Marcel Dekker, Inc.   New York and Basel

Library of Congress Cataloging in Publication Data

Surfactant-based separation processes / edited by John F. Scamehorn,
   Jeffrey H. Harwell.
      p.    cm. — (Surfactant science series : v. 33)
      Includes bibliographies and index.
      ISBN 0-8247-7929-0 (alk. paper)
      1. Separation (Technology)   2. Surface active agents.
   I. Scamehorn, John F.  II. Harwell, Jeffrey H.
   III. Series.
   TP156.S45S875  1989                        88-35241
   660.2'842—dc19                             CIP

This book is printed on acid-free paper.

MARCEL DEKKER, INC.
270 Madison Avenue, New York, New York   10016

Current printing (last digit):
10  9  8  7  6  5  4  3  2  1

PRINTED IN THE UNITED STATES OF AMERICA

*To Professor Robert S. Schechter who showed us the importance of personal and scientific integrity and introduced us to the profound lessons to be learned from the study of surfactants.*

# Preface

A new generation of industrial separation processes is emerging: surfactant-based separations. These new processes seem uniquely well suited to certain types of separation problems that are of growing importance in industry. Areas of application in which these new techniques will find increasing utilization include the following:

1. *Biotechnology.* Here the products that need to be recovered are generally found in dilute aqueous solution and are easily degraded. The mild conditions under which surfactants effect a separation protect valuable biochemically produced products.

2. *Pollution control.* Surfactants can be used to remove dissolved organics and heavy metals from wastewater with significantly less energy than traditional processes, and without introducing substantial toxicity from residual surfactants. Additionally, the final solutions produced are so concentrated that in many cases waste streams will be converted into new product streams.

3. *Energy conservation.* Surfactant-based separation processes can have a large advantage over traditional methods since they generally require little energy themselves and downstream process water often does not require further treatment.

This volume brings together articles on a number of surfactant-based separation techniques that the editors feel have the potential for wide industrial application. Although there had previously been a relatively low level of activity in this field, in the last decade there has been an explosion of new ideas for applying surfactants to separations. A number of the chapters in this book therefore concern processes that are yet to be commercialized. Some chapters are essentially original research papers about totally new methods. The astute industrial researcher may well find a new technique in this book that solves separation problems of immediate concern.

Both academic and industrial researchers will recognize that the techniques discussed here raise a number of interesting fundamental questions concerning both separation science and colloid and surface science. There will almost certainly be an increase in research in both academia and industry on surfactant-based separation processes in the future.

The purposes of this book are, then, to summarize the most important current surfactant-based separations for the consideration of industrial workers and to alert both academic and industrial researchers to this promising new field. We hope that new permutations of current ideas can be considered, totally new processes invented, and the underlying science further elucidated.

John F. Scamehorn<br>
Jeffrey H. Harwell

# Contributors

**D. Lowry Blakeburn***   Institute for Applied Surfactant Research, University of Oklahoma, Norman, Oklahoma

**Lori L. Brant***  Institute for Applied Surfactant Research, University of Oklahoma, Norman, Oklahoma

**Thomas E. Carleson**   Department of Chemical Engineering, University of Idaho, Moscow, Idaho

**Sherril D. Christian**   Institute for Applied Surfactant Research, University of Oklahoma, Norman, Oklahoma

**Rex T. Ellington**   Institute for Applied Surfactant Research, University of Oklahoma, Norman, Oklahoma

**Stig E. Friberg†**   Department of Chemistry, University of Missouri-Rolla, Rolla, Missouri

**Douglas W. Fuerstenau**   Department of Materials Science and Engineering, University of California at Berkeley, Berkeley, California

**Nancy D. Gullickson**   Institute for Applied Surfactant Research, University of Oklahoma, Norman, Oklahoma

*Present affiliations:*
*Conoco Inc., Ponca City, Oklahoma
†Amoco Production Company, Denver, Colorado.
‡Chemistry Department, Clarkson University, Potsdam, New York.

Jeffrey H. Harwell*    Institute for Applied Surfactant Research, University of Oklahoma, Norman, Oklahoma

T. Alan Hatton    Department of Chemical Engineering, Massachusetts Institute of Technology, Cambridge, Massachusetts

Ronaldo Herrera-Urbina[†]    Department of Materials Science and Engineering, University of California at Berkeley, Berkeley, California

Chonlin Lee    Institute for Applied Surfactant Research, University of Oklahoma, Norman, Oklahoma

Parthasakha Neogi    Department of Chemical Engineering, University of Missouri at Rolla, Rolla, Missouri

Edgar A. O'Rear    School of Chemical Engineering and Materials Science, University of Oklahoma, Norman, Oklahoma

John F. Scamehorn    Institute for Applied Surfactant Research, University of Oklahoma, Norman, Oklahoma

Felix Sebba    Department of Chemical Engineering, Virginia Polytechnic Institute and State University, Blacksburg, Virginia

Kevin L. Stellner    Institute for Applied Surfactant Research, University of Oklahoma, Norman, Oklahoma

Jengyue Wu[‡]    Institute for Applied Surfactant Research, University of Oklahoma, Norman, Oklahoma

_Present affiliations_:
*(Through July, 1989) Directorate for Engineering, Division of Chemistry, Biochemistry, and Thermal Engineering, National Science Foundation, Washington, D.C.
[†]Saltillo Institute of Technology, Saltillo, Coahuila, Mexico.
[‡]Division of Polymer Science, Institute for Research in Chemical Industry, Hsinchu, Taiwan.

# Contents

# SURFACTANT-BASED
# SEPARATION PROCESSES

# I
# SEPARATIONS USING MEMBRANES

# 1

# Use of Micellar-Enhanced Ultrafiltration to Remove Dissolved Organics from Aqueous Streams

SHERRIL D. CHRISTIAN and JOHN F. SCAMEHORN    Institute
for Applied Surfactant Research, University of Oklahoma,
Norman, Oklahoma

The authors appreciate the financial support of the Office of
Basic Energy Sciences, Department of Energy, Grant No.
DE-FG05-84ER13678, and Department of Energy Grant No. DE-
FG01-87FE61146. Support has also been provided by the Okla-
homa Mining and Minerals Resources Research Institute and the
University of Oklahoma Energy Research Institute.

## SYNOPSIS

Micellar-enhanced ultrafiltration (MEUF) is an effective method
for removing dissolved organic solutes from aqueous streams.
An appropriate concentration of surfactant is added to a stream
containing organic pollutants or target compounds, so that a
large fraction of the surfactant exists in micellar form.  When the
resulting solution is passed through an ultrafilter having pore
diameters smaller than the micelle diameter, most of the surfactant
and the organic solute remain in micelles in the retentate solution.
The permeate solution passing through the membrane is in many
cases practically pure water.
     The conditions under which MEUF is effective in removing
organics are described, including the choice of membrane molecular weight cutoff values, flow rates, and concentration ranges
for surfactants and solutes.  Frequently, excellent predictions
of the effectiveness of MEUF in removing organic solutes can be
made from knowledge of equilibrium solubilization isotherms, obtained by a variety of experimental techniques.  A group contribution model is described for predicting the extent of solubilization of organic solutes in micelles at low solute concentrations;
the method is employed to predict the efficiency of MEUF
separations.

## I.  INTRODUCTION

Micellar-enhanced ultrafiltration (MEUF) has been shown to be
an effective method for removing either dissolved organic compounds or polyvalent ions from aqueous streams (1-9).  Recent
studies also indicate that the method can be used to remove
divalent ions and organic solutes simultaneously (8).  The present
chapter considers the removal of organic solutes from water and

Chapter 2 discusses the removal of multivalent ions from water using MEUF.

The basic process used to remove organic solutes from aqueous streams with MEUF can be described in relation to the drawings in Fig. 1. In typical applications, an ionic or nonionic surfactant is added to the pollutant stream at concentrations well above the critical micelle concentration (CMC), so that most of the added surfactant exists in micellar form. Frequently, micelles contain 50–150 molecules of surfactant bound together in spherical or spheroidal aggregates. When the polluted stream containing the added surfactant is forced through an ultrafilter, most of the surfactant remains in the retentate solution *provided* the pores of the ultrafilter have diameters smaller than the micelles. The successful removal or organic pollutants can be accomplished if a significant fraction of these solutes resides within the micelles.

Concentration polarization, as well as diffusional effects relating to the sizes of low molecular weight solutes, can influence the relative concentrations of these species in the permeate solutions in MEUF experiments (2,5). Nonetheless, in numerous studies, it has been shown that the results of MEUF separations can be predicted with good accuracy by assuming that the concentrations of small solute molecules in the permeate solution are equal to the concentrations of these species in free or monomeric form in the retentate solution (2,3,5). Thus, in predicting MEUF separation efficiencies, it is useful to have accurate information about the *equilibrium* solubilization behavior of particular organic solubilizate species.

The present chapter begins with a summary of experimental results that provide the basis for using MEUF in a wide range of separations (3,5). Extensive separation results have already been obtained, employing the surfactant hexadecylpyridinium chloride or cetylpyridinium chloride (CPC) and using membranes having pore diameters in the range 1,000- to 50,000-dalton molecular weight cutoff (MWCO). Studies have been made of variations of flux through the ultrafiltration membrane as a function of retentate concentrations of CPC, varying from 15 mM practically up to the gel point (approximately 530 mM CPC). Experimental solubilization results are presented for typical organic solutes in aqueous surfactant solutions; consideration is given to alternative ways of processing solubilization data to provide the information needed in planning MEUF separations. Data are reported for a number of typical solutes in micelles of CPC and sodium dodecyl sulfate (SDS). To simplify the planning of MEUF separations of solutes for which solubilization data have not been

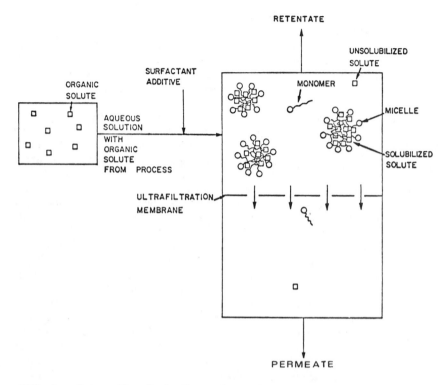

FIG. 1   Schematic of micellar-enhanced ultrafiltration to remove dissolved organics from water.

obtained, we describe a recently proposed group contribution method for predicting the extent of solubilization of dilute organic solutes by CPC and SDS. Finally, calculations are given to show how effective MEUF can be in reducing the concentrations of several typical organic solutes assumed to be present simultaneously, at small concentrations, in a wastewater stream.

## II.  PERFORMANCE OF MICELLAR-ENHANCED ULTRAFILTRATION IN REMOVING AN ORGANIC SOLUTE FROM WATER

Figure 1 indicates the simple principles that operate to make MEUF an efficient separation method.   An aqueous stream, containing an organic solute, is treated with a surfactant to produce a

micellar solution in which the organic compound is mostly solubilized or bound by the micelles. Ultrafiltration of this solution
produces a permeate stream that can have very low concentrations
of both the organic solute and the surfactant. In the following
sections of this chapter, solubilization results will be considered
for a number of typical organic solutes. However, it is informative first to consider in some detail results for a single organic
solute, 4-tert-butylphenol (TBP), removed by ultrafiltration of
an aqueous stream to which CPC has been added (3,5).

Figures 2 and 3 show MEUF results at 30°C for aqueous
streams containing TBP and CPC, obtained for solutions having
mole ratios of TBP to CPC of 1:10 for various membranes (Fig. 2)
and both 1:10 and 1:20 (paths 1 and 2, respectively, in Fig. 3).
The results in Fig. 2 indicate that for all of the ultrafiltration
membranes having MWCO values less than 50,000, the concentration of CPC appearing in the permeate solution is less than the
mean ionic molality of CPC (10), which is slightly greater than
the critical micelle concentration of the surfactant (0.88 mM).
Thus, hindrance of the CPC monomers reduces the concentration
of surfactant transferred from the retentate to the permeate to
values less than those expected at equilibrium. The fact that so
little CPC passes through the membrane is very good evidence
that the micelles are quantitatively excluded by pores having
MWCO values less than 50,000.

At retentate concentrations of CPC in excess of about 200 mM,
sizable increases occur in the permeate concentrations of CPC.
This suggests that submicellar aggregates containing several CPC
ions and solubilized TBP molecules form in the permeate; these
complexes are apparently small enough to pass through the pores
of the membrane.

Figure 3 shows the effectiveness of MEUF in removing TBP
from mixtures of TBP and CPC having 1:10 and 1:20 mole ratios
(paths 1 and 2). At concentrations of CPC less than 230 mM on
either path, the permeate TBP concentration nearly equals that
of the unsolubilized TBP in the corresponding retentate solution,
inferred from equilibrium solubilization experiments (3). The
concentration of TBP in the permeate solution remains nearly
constant in this low-concentration region, equaling about 0.08 mM
along path 1 and 0.04 mM along path 2. As would be expected
for a constant partition coefficient for the organic solute between
the bulk aqueous solution and the micelles, doubling the ratio of
surfactant to solute in the retentate halves the concentration of
solute in the permeate. The observed permeate concentrations
are equivalent to rejections of TBP varying from 97.6 to 99.7%

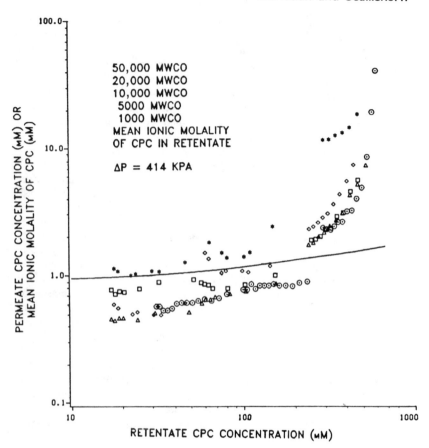

FIG. 2 Concentration of CPC in permeate and CPC mean ionic molality for membranes of various pore sizes.

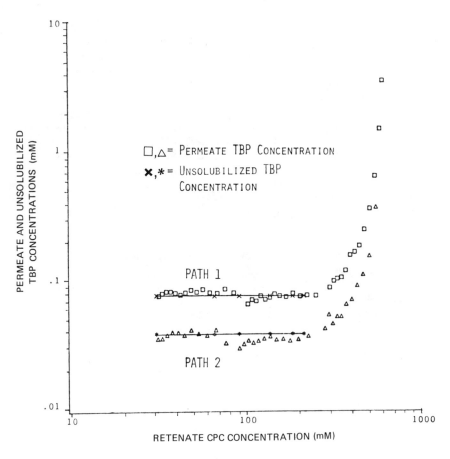

FIG. 3   TBP concentration in permeate for different TBP/CPC ratios in retentate or "paths."

along the paths shown in Fig. 3. {Rejection (in %) is defined
as 100·[1-(permeate)/(retentate)].} The rapid rise in concentration of TBP in the permeate, at CPC concentrations in the
retentate greater than about 220 mM, parallels the increase in
permeate CPC concentrations (noted above) for the same solutions.

The results in Figs. 2 and 3 show that at CPC concentrations
greater than about 200 mM, MEUF becomes less effective in separating the organic solute from an aqueous stream because relatively high concentrations of both CPC and TBP penetrate the
membrane. It is also useful to consider flux results for MEUF
experiments performed at varying CPC concentrations, both
below and above 200 mM. Figure 4 depicts flux data obtained
for retentate CPC concentrations varying from about 15 mM to
greater than 500 mM. Independent of the MWCO values of the
different membranes, at the higher CPC concentrations the flux
tends nearly linearly toward zero on each of the plots of flux
vs. log CPC concentration. The CPC concentration in the
retentate at which the extrapolated flux reaches zero (the so-
called gel point) is approximately 530 mM.

The fact that the gel point lies in a range of CPC concentrations considerably greater than that which is optimal for removing the organic solute implies that flux limitations do not re-
strict the use of CPC in MEUF separations. It is important to
note that the fluxes are not substantially reduced below those of
pure water (with the same membrane) if retentate CPC concentrations are kept below 200 mM. Figure 4 shows that the flux
is considerably increased by using membranes with MWCO values
in the range 20,000–50,000. The enhancement in flow rates may
justify using these membranes in separations, although rejection
of both the surfactant and the organic solute is slightly poorer
in the case of the larger MWCO membranes.

Comprehensive data similar to those in Figs. 2–4 have only
been obtained for a few organic solutes and two ionic surfactants [CPC and SDS (7), which has a gel point of approximately 570 mM (see Chapter 2)]. Nonetheless, it is reasonable
to believe that the simple behavior described for the removal of
TBP with the surfactant CPC will occur for a wide range of
solutes and surfactants. In many MEUF applications, aqueous
streams will have mole ratios of organic to surfactant even smaller
than 1:10. Consequently, there is little reason to anticipate that
diminished fluxes or a rapidly increasing loss of surfactant and
organic through the membrane will cause problems in MEUF separations at CPC or SDS concentrations varying from low values
up to at least 200 mM.

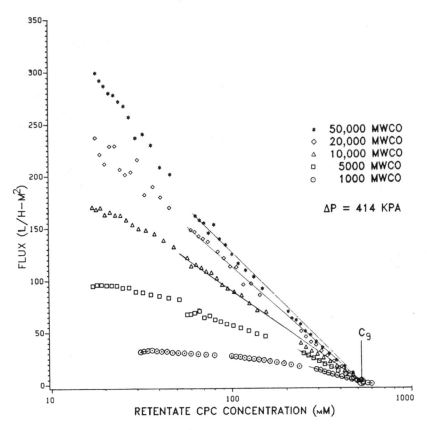

FIG. 4   Flux for various membrane pore sizes.

Although our use of MEUF has been restricted primarily to
CPC, SDS, and a few other ionic surfactants, there is no reason
why numerous commercially available surfactants could not be
used in MEUF separations.   When ionic surfactants are to be
employed, separations should not be planned for temperatures
lower than the Krafft temperature, and with nonionic surfactants,
problems may occur with phase separation at temperatures above
the cloud point.   Membrane MWCO values in the range 10,000–
50,000 should be suitable in many applications.   In practical
applications of MEUF, one needs to balance the possibility of
attaining greater fluxes with membranes having larger pore

diameters against the fact that somewhat better separation efficiencies (larger rejections) can be obtained with membranes
having smaller pores.

It should be emphasized again that in separating organic
solutes (either singly or in a complex mixture) from aqueous
streams, *information about solubilization equilibrium behavior
alone* can be used to predict MEUF rejections. Therefore, it becomes very important to have as complete information as possible
about the extent of solubilization of particular organic compounds
by surfactants or surfactant mixtures. The following sections
of this chapter describe the phenomenon of solubilization of
organic solutes by surfactant micelles and outline a practical
method for predicting equilibrium solubilization behavior.


## III.  SOLUBILIZATION ISOTHERMS

Numerous types of experiments have been used to infer the extent
of solubilization of organic solutes by aqueous surfactant solutions;
among the many available techniques are vapor pressure (11,12),
gas or vapor solubility (13,14), nmr (15), membrane (16,17),
and phase equilibrium (liquid solubility) (18) methods.  No attempt
will be made to review the advantages or disadvantages of the
several techniques; a review article provides critical compar-
isons of the available solubilization methods (19).  We have emphasized the importance, in planning surfactant-based separations,
of utilizing physical methods that provide solubilization results
throughout wide ranges of intramicellar composition.  Unfortunately, one of the simplest solubilization methods, the so-called
maximum additivity or maximum solubility method (18), yields
only a single value of the total concentration of an organic compound (at saturation) in equilibrium with the micellar solution.
In using MEUF, the concentration of unsolubilized organic compound is ordinarily much smaller than saturation; therefore, such
information is usually not adequate for predicting the efficiency
of MEUF separations.

Our work has shown that vapor-pressure methods, when
applicable, are capable of yielding the most highly precise
solubilization isotherms for organic solutes in aqueous surfactant
systems.  For solutes not sufficiently volatile to be investigated
by vapor pressure measurements, we have found that a new
membrane method [semiequilibrium dialysis, SED (16,17,20,21)]
is convenient for determining the extent of solubilization as a
function of the intramicellar mole fraction of organic solute $X_O$.

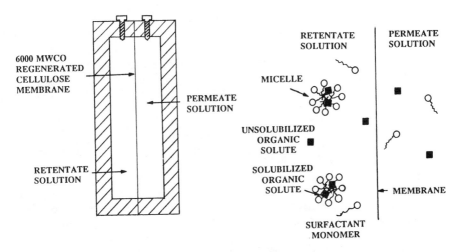

FIG. 5 Equilibrium dialysis cell and distribution of solute and surfactant species at semiequilibrium.

The diagram in Fig. 5 is useful in explaining how SED may be used to infer equilibrium solubilization results. We have called the method "semiequilibrium" dialysis because only the organic solute (but not the surfactant) reaches equilibrium with respect to both the permeate and the retentate solutions. The surfactant continues to diffuse through the membrane for long periods of time, although within a 24-hr period its concentration in the permeate compartment usually does not greatly exceed the CMC.

In performing SED experiments, a solution of organic solute and surfactant is placed initially in the retentate compartment of the SED cell; pure water (or an electrolyte solution) is placed in the permeate compartment. After 16–24 hr, small concentrations of the surfactant will have diffused through the membrane (which is typically a regenerated cellulose membrane having a MWCO value of 6,000). The organic solute transfers through the membrane more rapidly than monomers of the surfactant, and after about 16 hr it will have reached equilibrium with respect to both the permeate and the retentate solutions. Fortunately, within this time period, the concentration of micelles in the permeate compartment remains quite small, so that one may ordinarily assume that the total concentration of organic solute in the permeate compartment (determined after approximately 1 day) is equal to the concentration of the free organic solute. Therefore, it may be assumed that the analytical concentration of the organic species in the permeate equals the concentration of the mono- meric organic solute in the retentate compartment (16).

By making the assumption that the organic solute is at equi- librium with the solutions on both sides of the dialysis membrane, one can use SED results to calculate the fraction of the organic solute molecules in the retentate solution that is solubilized with- in the micelles and the fraction that remains unsolubilized. In the most accurate studies, corrections are made to account for the solubilization of organic solutes by the small concentration of micelles in the permeate solution (17,20).

For many purposes, including most treatments of solubilization data, it is convenient to assume that aqueous solutions of sur- factants may be treated as consisting of a micellar pseudophase and the bulk aqueous solution outside the micelles. To be sure, elaborate mass action models have sometimes been used to account for equilibria between monomers of the surfactant (and any counterions that may be present) and the micellar aggregates, which are assumed to have various stoichiometries. Although it may be argued that the mass action models are fundamentally more nearly correct, it is quite difficult to apply them even to binary aqueous solutions in which large micelles form. There- fore, it not surprising that few attempts have been made to use mass action models to correlate data for ternary aqueous systems containing variable amounts of the organic solute, solubilized within large micellar species.

Precise vapor pressure and membrane equilibrium studies of solubilization have provided justification for using the pseudo- phase equilibrium assumption instead of the more complicated mass action model. It is frequently observed that solubilization iso- therms, relating either the activity coefficient of the organic compound in the intramicellar solution or the solubilization equilib- rium constant to the intramicellar mole fraction of organic solute $X_O$, do not depend on the total concentration of surfactant (17,20,21). Thus, in studies with SDS and CPC, the total sur- factant concentration can ordinarily be varied from very small values to 0.2 or 0.3 M *without* substantially altering the solubil- ization isotherm. This observation is consistent with the pseudo- phase equilibrium model. At concentrations of SDS or CPC in excess of 0.4-0.5 M, the isotherms do change, probably reflecting a transition of the micelles from the spherical forms thought to be present at low concentrations to rod-shaped aggregates.

Additional justification for applying the pseudophase equilibrium model in solubilization studies has been provided by using a Gibbs-Duhem integration of solute activity data (consistent with the pseudophase equilibrium model) to predict surfactant activities for ternary solutions containing benzene and sodium octyl sulfate [22]. Values of the surfactant activities obtained in this way are in excellent agreement with activities inferred by applying a rather elaborate mass action model to the same results. In systems containing much larger micelles (such as the SDS and CPC solutions under consideration here), the pseudophase assumption should be an even better approximation. In any event, the pseudophase model has been applied to all of the results reported in this chapter.

If the CMC of the surfactant is quite small compared to the total concentration of surfactant in solution, it is justifiable to assume that most of the surfactant exists in micellar form. However, in the most precise studies, corrections have been made to account for the concentration of surfactant present as the monomer. In the case of organic compounds that are moderately soluble in pure water, it is also necessary to estimate the concentration of monomeric solute in the "bulk aqueous solution"; i.e., dissolved in the part of the solution excluding the micellar species. Methods for inferring how both the surfactant and the organic solute are partitioned between the intramicellar solution and the bulk aqueous phase will not be discussed here. It will be assumed that the mole fraction of organic compound contained in the micellar species ($X_O$) has been correctly determined from equilibrium solubilization studies, along with values of the concentration of the organic compound present as the monomeric species in the bulk aqueous phase ($c_O$). Most investigators who present quantitative solubilization results provide data comprising sets of $X_O$, $c_O$ values, or at least give results that can be converted into this form.

## IV. SOLUBILIZATION EQUILIBRIUM CONSTANTS AND ACTIVITY COEFFICIENTS FOR SOLUBILIZED SPECIES

Several choices may be made as regards the mathematical form of mass action equilibrium constants, solubilization constants, partition coefficients, and other parameters that are used to represent solubilization data. In our research, we have preferred to use an apparent solubilization constant, defined by

$$K = \frac{X_O}{c_O} \qquad\qquad (1)$$

to describe data, where $c_O$ is the concentration of free or monomeric organic solute in the bulk aqueous phase.  Plots of K vs.
$X_O$ are referred to as solubilization isotherms.  K pertains to
the equilibrium

Organic solute + surfactant micelle = organic solute · micelle

and in the limit as $X_O$ approaches 0, K is the equilibrium constant (in reciprocal molarity units) for transferring a single
molecule of the organic solute from its ideally dilute solution in
the bulk aqueous phase into the micelle.  If the micelle aggregation
number n is known, it is possible to multiply K by this number to
obtain the mass action equilibrium constant for the reaction

Organic solute + $A_n$ = organic solute · $A_n$

which represents the solubilization of one molecule of the organic
solute by a single micelle $A_n$.  In this limit, it is usually reasonable to assume that the solubilization of one molecule of the organic solute does not affect the micelle aggregation number n.
Knowledge of K as a function of $X_O$ makes it possible to calculate
the concentration of free organic solute in the bulk aqueous solution at given total concentrations of the surfactant and the organic
compound in the retentate solution in a MEUF experiment.  By
making the previously justified assumption that the permeate
solution contains the same concentration of free organic solute as
the retentate, it is thus possible to predict the efficiency of
MEUF separations.  When greater accuracy is called for, corrections may be made for the "salting-out" effect of an ionic surfactant solution on the concentration of the free organic compound
in the retentate phase.
     Other workers have defined the partition coefficient as the
ratio of the molar concentration of organic solute in the micellar
pseudophase to the molar concentration of the (presumably) unassociated organic compound in the bulk phase (23), or as the
ratio of the mole fractions of the solute in the two regions (24,25).
Given information about densities of the aqueous surfactant/or-
ganic solute systems, it is possible to interconvert values of solubilization constants of these various types.

Figure 6 includes plots of K vs. $X_O$ (the intramicellar mole fraction of the hydrocarbon solute) for solutions of cyclohexane in sodium dodecyl sulfate (SDS) at 25, 35, and 45°C. Included in the upper half of the figure are values of the rational activity coefficient of cyclohexane ($\gamma_O$) also plotted against $X_O$. $\gamma_O$ is defined as the ratio of the fugacity of the solute ($f_O$) to the fugacity of the pure liquid organic compound ($f_O^0$) at the same temperature divided by $X_O$. Values of K are related to the $\gamma_O$ values by the equation

$$\gamma_O = \frac{1}{Kc_O^0} \tag{2}$$

where $c_O^0$ represents the (hypothetical) molar concentration of monomeric organic solute in the bulk aqueous soltuion at which the fugacity of the solute would equal that of the pure liquid organic compound, using an extrapolation of the limiting Henry's law relationship between $f_O$ and $c_O$. In the case of liquid solutes that are only slightly soluble in water, $c_O^0$ is ordinarily nearly equal to the solubility of the compound in the bulk aqueous solution. In any event, K values may be readily converted into $\gamma_O$ values if the Henry's law constant for the organic solute, at infinite dilution in the bulk aqueous solution, is known.

Although it is not necessary to recast the solubilization results in the form of the $\gamma_O$ vs. $X_O$ curves, there are several reasons why it is informative to do so. First of all, $\gamma_O$ values provide a direct indication of the excess free energy of transfer of an organic solute from the pure liquid phase into the micelle at any given value of $X_O$. Thus, the fact that the value of $\gamma_O$ for cyclohexane in CPC, at small values of $X_O$, is greater than 5 shows that the free energy of transfer of cyclohexane from its pure liquid phase into the micelle is moderately positive. As $X_O$ increases, the value of $\gamma_O$ decreases, up to the point at which the micellar pseudophase becomes saturated with cyclohexane, and one anticipates that $\gamma_O$ will tend to approach unity more closely as $X_O$ increases toward unity.

By examining the shape of a plot of $\gamma_O$ vs. $X_O$ for a given solute, one can determine immediately whether the intramicellar environment at any value of $X_O$ is more or less "favorable" (in the Gibbs free energy sense) than the pure liquid phase of the organic compound. Compounds that are strongly solubilized (such as aliphatic alcohols and carboxylic acids) have values of $\gamma_O$ considerably smaller than unity, while those that are poorly

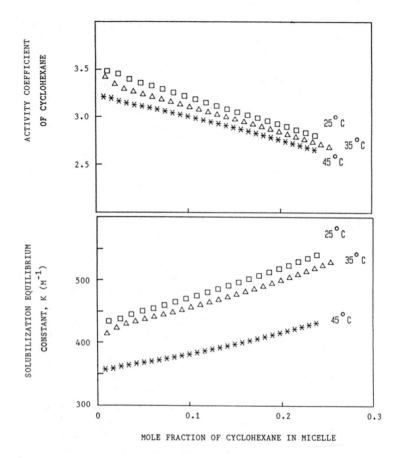

FIG. 6  Activity coefficients and solubilization equilibrium constants for cyclohexane in micelles of 0.1 M CPC.

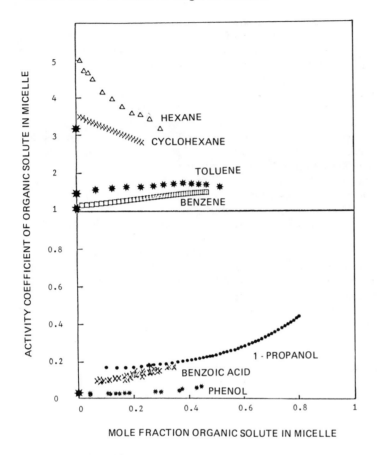

FIG. 7   Activity coefficients for organic solutes in micelles of
0.1 M CPC at 25°C.

solubilized (in comparison with the pure liquid phase as a reference) will have large values of $\gamma_O$.

Figure 7 shows plots of $\gamma_O$ vs. $X_O$ for several organic solutes in CPC intramicellar solutions. The tendency of $\gamma_O$ to approach unity as $X_O$ increases to sufficiently large values is observed in each case. Moreover, the compounds are conveniently divided into two classes: the highly polar molecules, which have $\gamma_O$ values considerably less than unity, and for which $\gamma_O$ almost always increases as $X_O$ increases; and the apolar aliphatic solutes, which have $\gamma_O$ values greater than unity. In the intermediate region are solutes like benzene and toluene, which have values of $\gamma_O$ closer to unity and for which it can be argued that solubilization occurs to a significant extent both in the hydrophobic micellar interior and in the vicinity of the polar-ionic surface region of the micelles. Although many recent studies from this laboratory have provided extensive solubilization data (in the form $\gamma_O$ vs. $X_O$ or K vs. $X_O$), the typical features of solubilization isotherms inferred for a wide range of solute types are fairly represented by the results in Fig. 7. A recent doctoral dissertation (21) contains experimental data for many of the systems described in this chapter as well as extensive results for numerous other aqueous solutions of surfactants and surfactant mixtures.

When $\gamma_O$ vs. $X_O$ curves are available for numerous solutes in a given surfactant, it is relatively easy to predict approximate activity coefficient isotherms for structurally similar compounds for which solubilization data are not available. Then $\gamma_O$ values for any given solute can be converted into the K values needed to predict MEUF separation factors, *provided* that Henry's law constants are available for the solute in the bulk aqueous phase. For slightly soluble organic liquid solutes, the conversion from $\gamma_O$ values to K values is even simpler. One can directly use the relation $K = 1/(\gamma_O c_O^o)$, where $c_O^o$ is the solubility of the compound in water.

## V.  A GROUP CONTRIBUTION METHOD FOR PREDICTING SOLUBILIZATION EQUILIBRIA

Attempts have been made to predict solubilization equilibria for organic solutes, using both a group contribution approach (26) and a correlation that relates the partition coefficients of these solutes in the micelles to partition coefficients of the solutes between water and octanol (27,28).

Recently, we proposed a simple group contribution method that can be used to make good predictions of limiting solubilization equilibrium constants for solutes in SDS and CPC (29). The method considers the three equilibria involved in the cycle

solute [unit molarity, ideal gas]

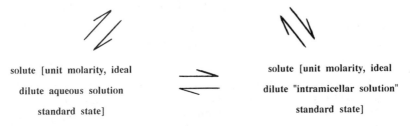

solute [unit molarity, ideal            solute [unit molarity, ideal

dilute aqueous solution                 dilute "intramicellar solution"

standard state]                         standard state]

where the equilibrium between the solute in dilute aqueous solution and the solute solubilized in the micelle is the one of primary interest. Experimental results are assumed to be available to calculate partition constants (equivalent to Henry's law constants) for transferring solutes from the dilute binary aqueous solution to the ideal gaseous phase; the group contribution model is used to predict the dimensionless partition coefficients, $K' = [solute]_{micelle}/[solute]_{gas\ phase}$. Therefore, by completing the cycle, values of the solubilization constant K or the activity coefficient $\gamma_O$ for organic solutes of interest can be estimated.

In using the model, values of K' are predicted from the equation

$$\ln K' = \sum_i n_i b_i \qquad (3)$$

where $n_i$ represents the number of groups of a particular type (e.g., number of carbons, number of hydrogens, number of aryl groups, number of hydroxyls) and $b_i$ represents the contribution to ln K' of the ith type of group. Table 1 includes values of the group factors [$b_i$ in Eq. (3)] useful for predicting K' values for solutes in CPC and SDS. Average deviations between calculated and experimental K' values (and hence deviations in K) are approximately 30% for the SDS systems and 15% for the CPC systems (29). Limiting values of K and $\gamma_O$, at $X_O = 0$, calculated with this method, are indicated by asterisks for several of the systems represented in Fig. 7.

TABLE 1   Group Parameters in Eq. (3) for SDS and CPC

|                     | SDS            | CPC            |
| ------------------- | -------------- | -------------- |
| For each C atom     | 2.14 ± 0.30    | 2.14 ± 0.30    |
| For each aryl group | −3.22 ± 1.12   | −2.68 ± 1.06   |
| For each H atom     | −0.64 ± 0.13   | −0.65 ± 0.13   |
| For each OH group   | 8.57 ± 0.22    | 8.91 ± 0.18    |

Although predictions of the group contribution model need to be tested against results for a wide range of surfactant-organic solute systems, the method is attractive because it provides estimates of values of K for organic compounds from knowledge of solute structural formulas and limiting Henry's law constants alone.   Henry's law constants are available in the literature for many of the common pollutants and other important target compounds in aqueous streams (30,31).   However, where these constants are not known, they may readily be estimated (for any sparingly soluble compound) from values of the vapor pressure of the compound and its solubility ($c^0_O$) in water.   Thus, for the solute benzene, which has a solubility of 0.022 M or 0.00040 m.f. at 25°C (32) and a vapor pressure of 95.18 torr (33), we estimate that the Henry's law constant is 95.18/0.00040 = $2.4 \times 10^5$ torr m.f.$^{-1}$.   The most precise study of the partial pressure of benzene as a function of its mole fraction in water reported a value of the limiting Henry's law constant at 25°C of $2.36 \times 10^5$ torr m.f.$^{-1}$ (34).

## VI.   MICELLAR-ENHANCED ULTRAFILTRATION OF TYPICAL ORGANIC SOLUTES IN A WASTEWATER STREAM

Our previous experience with the MEUF method indicates that equilibrium solubilization results can ordinarily be used to make reliable predictions of separation efficiencies.   Results of this type of calculation may be illustrated for a hypothetical wastewater stream containing small concentrations of six organic solutes present simultaneously.   The assumed retentate solution contains initially 50 mM CPC and 0.5 mM each of the solutes benzene, cyclohexane, phenol, p-cresol, 1-pentanol, and chlorobenzene.

TABLE 2  Performance of MEUF in Removing Organic Solutes from an Aqueous Stream[a]

| Solute | $K(M^{-1})$[b] | Final concentration (mM) | | Rejection (%) | |
|---|---|---|---|---|---|
| | | Permeate | Retentate | Calculated from experimental K | Calculated from group contribution model (Ref. 29) |
| Benzene | 40 | 0.165 | 1.84 | 91.02 | 92.0 |
| Cyclohexane | 430 | 0.0219 | 2.41 | 99.09 | 99.2 |
| Phenol | 62 | 0.121 | 2.02 | 94.02 | 94.0 |
| p-Cresol | 195 | 0.0460 | 2.32 | 98.02 | 97.7 |
| 1-Pentanol | 15 | 0.285 | 1.36 | 79.15 | 81.8 |
| Chlorobenzene | 162 | 0.0543 | 2.28 | 97.63 | — |

[a]Feed: 0.5 mM of each organic solute; [CPC] = 50 mM. Permeate/feed = 0.8.
[b]Solubilization equilibrium constants from Ref. 21.

The solution is to be ultrafiltered until 80% of the volume of the solution is removed as permeate, i.e., the water recovery is 80%.

The results in Table 2 summarize the performance of MEUF in removing the six organic solutes under the assumed conditions. Values of the solubilization equilibrium constant K are given for the compounds along with the expected values of the concentrations of the various solutes in the permeate and retentate solutions. Also included in the table for each compound are values of the % rejection. For the overall mixture, the mole fraction of total organic solute in the micelle ($X_O$) is only about 0.06; in the calculations made to predict MEUF results, it has been assumed that $X_O$ is small enough that limiting K values from Ref. 21 (corresponding to $X_O = 0$) can be used throughout. The rejection percentages enclosed in parentheses are values calculated from the group contribution model using parameters from Table 1 and Henry's law constants derived from literature values of vapor pressure and solubility. The results in Table 2 show that good separations can be obtained for all six organic solutes in a single pass with the MEUF method.

## VII.  GENERAL DISCUSSION

In general, detailed solubilization isotherms are required to predict the effectiveness of MEUF in removing dissolved organic solutes in concentrated wastewater streams. Such results may be obtained from several physical methods that are capable of determining the dependence of the solubilization equilibrium constant K on the intramicellar mole fraction of the dissolved organic compound.

Frequently, the total concentration of organic solutes and the concentration of any single solute species in wastewater streams are so small that limiting values of the solubilization equilibrium constants (corresponding to $X_O = 0$) can be used in predicting MEUF results. In this limit, it is feasible to use either the extrapolated results of solubilization experiments (e.g., vapor pressure or SED studies performed throughout a range of values of $X_O$) or predicted limiting K values obtained with a group contibution model. Model predictions for numerous organic solutes, in either SDS or CPC micelles, have been shown to be in good agreement with experiment, so that reliable estimates can be made of the performance of MEUF experiments by using only information about the structural formulas of the organic solutes and their Henry's law constants in dilute aqueous solution.

TABLE 3  Removal of Selected Organic Solutes by CPC and SDS Micelles [a,b]

| Solute | Surfactant: CPC | | Solute | Surfactant: SDS | |
|---|---|---|---|---|---|
| | $K$ ($M^{-1}$) in CPC micelles | Rejection (%) | | $K$ ($M^{-1}$) in SDS micelles | Rejection (%) |
| Toluene | 121 | 97.0 | Benzene | 21 | 84.8 |
| 1-Butanol | 6 | 61.5 | Cyclohexane | 270 | 98.6 |
| 1-Hexanol | 30 | 88.8 | Phenol | 15 | 79.9 |
| o-Cresol | 187 | 98.0 | Benzyl alcohol | 13 | 77.5 |
| m-Cresol | 190 | 98.1 | 4-Tert-Butylphenol | 365 | 99.0 |
| Fluorobenzene | 52 | 93.3 | o-Cresol | 26 | 87.3 |
| o-Chlorophenol | 360 | 98.9 | Benzoic acid | 77 | 95.4 |
| p-Chlorophenol | 495 | 99.2 | Hexane | 630 | 99.4 |

[a]Feed: 2 mM of individual solutes; [CPC] or [SDS] = 50 mM.  Permeate/feed = 0.8.
[b]Calculations based on limiting K values at [organic solute] = 0, inferred from data in Ref. 21.

For many organic solutes, rejections in the range 95–99% are readily attainable under reasonable operating conditions in a single-pass MEUF process. The rejection of any given compound is favored by a large value of the solubilization equilibrium constant. Highly hydrophobic compounds, which have low solubilities in water, will have very large K values and will be almost quantitatively removed from wastewater streams by MEUF; compounds that are more soluble will ordinarily be less effectively removed in single-pass ultrafiltration processes.

Table 3 lists solubilization equilibrium constant values and calculated rejections for a number of solutes in CPC, computed by assuming that the aqueous stream contains 2 mM of the given solute and 50 mM CPC, and that water recovery is 80%. Even in the case of solutes not represented in Table 3, it may be feasible to estimate rejections for MEUF separations by extrapolating results listed in the table for similar compounds.


## ACKNOWLEDGMENT

Professor E. Tucker's contributions to solubilization theory and experimentation are greatly appreciated.


## REFERENCES

1.  P. S. Leung, in *Ultrafiltration Membranes and Applications*
    (A. R. Cooper, ed.), Plenum Press, New York, 1979,
    p. 415.
2.  J. F. Scamehorn and J. H. Harwell, in *Surfactants in Chemical/Process Engineering* (D. T. Wasan, M. E. Ginn and D. O.
    Shah, eds.), Marcel Dekker, New York, 1988, p. 77.
3.  R. O. Dunn, J. F. Scamehorn, and S. D. Christian,
    *Sep. Sci. Technol.*, 20:  257 (1985).
4.  L. L. Gibbs, J. F. Scamehorn, and S. D. Christian,
    *J. Membrane Sci.* 30: 67 (1987).
5.  R. O. Dunn, J. F. Scamehorn, and S. D. Christian,
    *Sep. Sci. Technol.*, 22: 763 (1987).
6.  S. N. Bhat, G. A. Smith, E. E. Tucker, S. D. Christian,
    and J. F. Scamehorn, *Ind. Eng. Chem. Res.*, 26: 1217
    (1987).
7.  D. A. El-Sayed, J. F. Scamehorn, and S. D. Christian,
    *Sep. Sci. Technol.* (in preparation).

8. R. O. Dunn, J. F. Scamehorn, and S. D. Christian, *Colloid Surf.* (submitted for publication).
9. G. A. Smith, S. D. Christian, E. E. Tucker, and J. F. Scamehorn, in *Ordered Media in Separations* (W. L. Hinze and D. W. Armstrong, eds.), ACS Symp. Ser., Vol. 342, 1987, p. 184.
10. D. S. Bushong, Ph. D. dissertation, University of Oklahoma, 1985,
11. E. E. Tucker and S. D. Christian, *Faraday Symp. Chem. Soc.*, *17*: 11 (1982).
12. E. E. Tucker and S. D. Christian, *J. Colloid Inter. Sci.*, *104*: 562 (1985).
13. I. B. C. Matheson and A. D. King, *J. Colloid Inter. Sci.*, *66*: 464 (1978).
14. A. Wishnia, *J. Phys. Chem.*, *67*: 2079 (1963).
15. P. Stilbs, *J. Colloid Inter. Sci.*, *87*: 385 (1982).
16. S. D. Christian, G. A. Smith, E. E. Tucker, and J. F. Scamehorn, *Langmuir*, *1*: 564 (1985).
17. G. A. Smith, S. D. Christian, E. E. Tucker, and J. F. Scamehorn, *J. Solution Chem.*, *15*: 519 (1986).
18. H. Saito and K. Shinoda, *J. Colloid Inter. Sci.*, *24*: 10 (1962).
19. C. M. Nguyen, J. F. Scamehorn, and S. D. Christian, *Tenside, Surfactants, Deterg.* (in press).
20. W. S. Higazy, F. Z. Mahmoud, A. A. Taha, and S. D. Christian, *J. Solution Chem.*, *17*: 191 (1988).
21. G. A. Smith, Ph. D dissertation, University of Oklahoma, 1986.
22. S. D. Christian, E. E. Tucker, G. A. Smith, and D. S. Bushong, *J. Colloid Inter. Sci.*, *10*: 439 (1986).
23. R. De Lisi, C. Genova, and V. T. Liveri, *J. Colloid Inter. Sci.*, *95*: 428 (1983).
24. A. M. Blokhus, H. Hoiland, and S. Backlund, *J. Colloid Inter. Sci.*, *114*: 9 (1966).
25. C. H. Spink and S. Colgan, *J. Phys. Chem.*, *87*: 888 (1983).
26. C. Hirose and L. Sepulveda, *J. Phys. Chem.*, *85*: 3689 (1981).
27. C. Treiner and A. K. Chattopadhyay, *J. Colloid Inter. Sci.*, *109*: 101 (1986).
28. C. Hansch and W. J. Dunn, *J. Pharm. Sci.*, *61*: 1 (1972).
29. G. A. Smith, S. D. Christian, E. E. Tucker, and J. F. Scamehorn, *Langmuir*, *3*: 598 (1987).

30.  A. W. Andren, W. J. Doucette, and R. M. Dickhut, in
     *Sources and Fates of Aquatic Pollutants* (R. A. Hites and
     S. J. Eisenreich, eds.), ACS Symp. Ser., Vol. 216, 1987, p. 3.
31.  I. Wichterle, J. Linek, and E. Hala, *Vapor-Liquid
     Equilibrium Data Bioliography*, Elsevier, Amsterdam, 1973;
     Supplements I–IV, 1976–1985.
32.  F. Franks, M. Gent, and H. H. Johnson, *J. Chem. Soc.,
     166*: 2716 (1963).
33.  R. C. Wilhoit and B. J. Zwolinski, *Handbook of Vapor
     Pressures and Heats of Vaporization of Hydrocarbons and
     Related Compounds.*  Thermodynamics Research Center,
     College Station, Texas, 1971.
34.  E. E. Tucker, E. H. Lane, and S. D. Christian, *J. Solution
     Chem., 10*: 1 (1981).

# 2

# Use of Micellar-Enhanced Ultrafiltration to Remove Multivalent Metal Ions from Aqueous Streams

JOHN F. SCAMEHORN, SHERRIL D. CHRISTIAN, and REX T. ELLINGTON    Institute for Applied Surfactant Research, University of Oklahoma, Norman, Oklahoma

Financial support for this work was provided by the Office of Basic Energy Sciences of the Department of Energy under Grant No. DE-FG05-84ER13678, the Department of Energy Grant No. DE-FG01-87FE61146, the Oklahoma Mining and Minerals Resources Research Institute, and the University of Oklahoma Energy Research Institute.

## SYNOPSIS

Micellar-enhanced ultrafiltration (MEUF) can be used to remove
dissolved multivalent ions from water.  Surfactant of opposite
charge to that of the ions is added to the aqueous stream.  The
surfactant micelles have a high electrical potential, causing the
multivalent ions to bind or adsorb on the micelles due to electrostatic attraction.  The stream can be treated by ultrafiltration
and the micelles rejected by large-pore membranes.  Rejections
of 99.8% have been observed for divalent metal ions with large
membrane pore sizes and high flux.  In this chapter, the effect
of system variables on performance of MEUF for removal of metals
and metal mixtures is discussed.  A theoretical model is outlined
for predicting rejections under a variety of condtions.

## I.  INTRODUCTION

Wastewaters containing dissolved metal ions provide a major
environmental hazard.  Sources of such wastewater include metal-
plating industries, circuit board manufacture, photographic and
photo-processing industries, synfuels plants, refineries, and
metal mine-tailing leachate.  Abandoned metal mines can fill with
water, the residual metal can be leached from the rock, and the
resultant water then enter the underlying aquifer or drain into
streams, polluting water supplies.  Improved technology has the
potential to reduce the concentration of metal ions emitted into
the environment and/or reduce the cost of water treatment.

Chapter 1 discussed the application of micellar-enhanced ultrafiltration (MEUF) to remove dissolved organics from water. This chapter discusses the removal of multivalent inorganic ions, such as heavy metals, from water. The basic physical mechanisms of removal are discussed and a mathematical model developed to describe the separation. Removal of metals and metal complexes having a variety of charges and structures, as well as removal of mixtures of metals, is described. Advantages of using surfactant mixtures are outlined and simultaneous removal of dissolved organics and metals is demonstrated.

## II.  BASIC PRINCIPLES

MEUF can be used to remove either cationic or anionic multivalent ions from water (1,2). Examples are divalent cationic copper or divalent anionic chromate ions. Surfactant is added to the aqueous stream containing the ion:  anionic surfactant for a cation; cationic surfactant for an anion. The surfactant forms micelles, roughly spherical aggregates containing generally 50–100 surfactant molecules. The micelles are highly charged and the ions of opposite charge to the surfactant (counterions) adsorb or bind onto the surface of the micelles. When these counterions are multivalent, the tendency to adsorb on the micelles is very great and only a small fraction of the counterion remains unbound under the appropriate conditions. As discussed in Chapter 1, nonionic organic solutes originally dissolved in the water will tend to solubilize in the interior of the micelles. Both the counterion binding and the organic solubilization can occur simultaneously.

The stream containing added surfactant is treated in an ultrafiltration unit with membrane pore sizes small enough to block the passage of the micelles with their bound counterion and solubilized organic. This process is shown schematically in Fig. 1. The nearly pure water passing through the membrane is the permeate and the concentrated stream not passing through the membrane is the retentate. The retentate is a waste stream if MEUF is removing pollutants from a wastewater stream; it is a product stream if the ion being removed is of value. In either case the retentate stream is much smaller in volume than the original feed water stream and is much less expensive to treat further or to dispose of. This chapter will focus on removal of multivalent ions, but simultaneous removal of ions and nonionic organic solutes will also be briefly discussed.

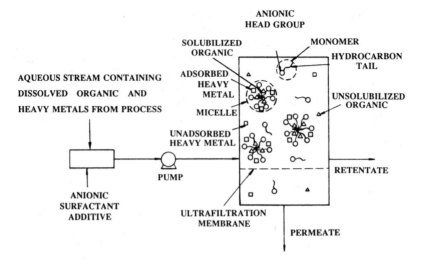

FIG. 1  Schematic of micellar-enhanced ultrafiltration to remove
dissolved heavy metal cations and organics from water.


## III. PERFORMANCE OF MICELLAR-ENHANCED ULTRAFILTRATION

In order to illustrate the efficiency of the process, the removal
of various divalent cationic metals using sodium dodecyl sulfate
(SDS) will be discussed.  The retentate-based rejection used
here is defined as

$$\text{Rejection (\%)} = 100 \left[1-\left(\left(\text{permeate[metal]}\right)/\left(\text{retentate[metal]}\right)\right)\right]$$

$$(1)$$

### A.  Effect of Metal Ion Concentration

The effect of varying the retentate divalent copper concentration
wile holding the retentate SDS concentration constant is shown in
Table 1 and Fig. 2.  At the lowest $Cu^{2+}$ concentration, the retentate is over 500 times as concentrated in copper as the permeate
(rejection = 99.82%).  A separation factor of 500 in one pass indicates the power of this new separation technique to remove metals
from water.  This also indicates that essentially all micelles are
being rejected, as will be further discussed later in this chapter.

TABLE 1    Results for SDS-Cu$^{2+}$ System

| Retentate concentration (mM) | | | Permeate concentration (mM) | Permeate rejection (%) | Membrane (MWCO) |
|---|---|---|---|---|---|
| SDS | Cu$^{2+}$ | NaCl | Cu$^{2+}$ | Cu$^{2+}$ | |
| 80 | 1.2 | 0 | 0.00220 | 99.82 | 1,000 |
| 80 | 4 | 0 | 0.0071 | 99.82 | 1,000 |
| 80 | 8 | 0 | 0.0374 | 99.53 | 1,000 |
| 80 | 16 | 0 | 0.257 | 98.39 | 1,000 |
| 80 | 32 | 0 | 2.16 | 93.25 | 1,000 |
| 40 | 8 | 0 | 0.117 | 98.54 | 1,000 |
| 160 | 8 | 0 | 0.0186 | 99.77 | 1,000 |
| 160 | 16 | 0 | 0.084 | 99.48 | 1,000 |
| 320 | 32 | 0 | 0.215 | 99.33 | 1,000 |
| 640 | 64 | 0 | 0.316 | 99.51 | 1,000 |
| 80 | 8 | 100 | 1.22 | 84.75 | 1,000 |
| 80 | 8 | 200 | 2.32 | 71.00 | 1,000 |
| 80 | 8 | 400 | 3.07 | 61.62 | 1,000 |
| 40 | 8 | 100 | 2.245 | 71.94 | 1,000 |
| 160 | 8 | 100 | 0.81 | 89.88 | 1,000 |
| 320 | 8 | 100 | 0.502 | 93.72 | 1,000 |
| 640 | 8 | 100 | 0.3073 | 96.16 | 1,000 |
| 80 | 8 | 0 | 0.0184 | 99.77 | 5,000 |
| 80 | 8 | 0 | 0.0247 | 99.69 | 10,000 |
| 80 | 8 | 0 | 0.0389 | 99.51 | 20,000 |

Temperature = 30°C; pressure drop = 414 kPa (60 psig).

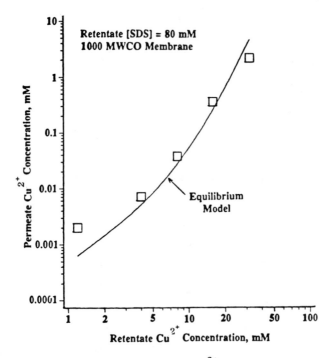

FIG. 2    Effect of retentate $Cu^{2+}$ concentration on permeate $Cu^{2+}$ concentration.

As the retentate $Cu^{2+}$ concentration increases, the permeate $Cu^{2+}$ concentration increases more than proportionally.  However, at the highest concentration used, the retentate $Cu^{2+}$ was 32 mM and the SDS concentration was 80 mM (the stoichiometric amount of $Cu^{2+}$ would correspond to approximately 36 mM).  Even at this nearly stoichiometric ratio, the binding of the copper by the micelles was so efficient that the rejection was 93.25%.

The improved percentage removal of copper at very low copper concentrations is a valuable practical aspects of MEUF. While some metal clean-up techniques, such as precipitation by pH adjustment, exhibit a decrease in efficiency as the metal becomes more dilute, MEUF exhibits an increase in efficiency.  As environmental regulations move to even lower allowable discharge levels, this could become increasingly important.

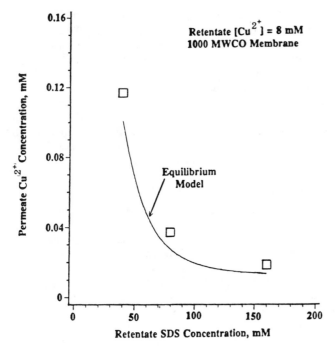

FIG. 3    Effect of retentate SDS concentration on permeate $Cu^{2+}$ concentration.

## B.  Effect of Surfactant Concentration

The minimum concentration of surfactant that is necessary to attain a desired separation is important to the economics of this process.   The effect of varying the surfactant concentration at constant retentate $Cu^{2+}$ concentration is shown in Table 1 and Fig. 3.   Increasing the surfactant concentration causes a decrease in the permeate $Cu^{2+}$ concentration.   The monomer concentration in equilibrium with these micelles is approximately 7 mM (1). One effect that occurs at low surfactant concentrations in the retentate is that a substantial fraction of the surfactant can be present as monomer (unassociated surfactant molecules) instead of in micellar form.   Therefore, at low retentate surfactant concentrations, the micellar surfactant concentration can increase in more than a proportional manner with total surfactant concentration. For example, for the conditions shown in Fig. 3, at a SDS concentration of 40 mM in the retentate, 88.1% of the surfactant is in micellar form, while at an SDS concentration of 160 mM in the retentate, 98.6% of the surfactant is in micellar form (1). This is partially responsible for the factor of 3 decrease in $Cu^{2+}$ concentration in the permeate that occurs when the surfactant concentration is doubled from 40 to 80 mM. At high surfactant concentrations, increasing the surfactant concentration does not cause the permeate metal concentration to decrease as rapidly as at lower surfactant concentrations.

## C. Effect of Retentate Concentration

During ultrafiltration, the vast majority of both the metal and the surfactant is rejected. Therefore, as an increasing amount of the original feed stream is emitted from the process as permeate, the retentate becomes more concentrated and the surfactant/multivalent ion ratio remains nearly constant in the retentate. This would correspond to increasing time of operation in a batch unit and to increasing residence time in the unit for a continuous membrane separation. In either case, if the permeate composition were known at a specific retentate composition, these data could be integrated along the path of the process to calculate the average permeate composition being emitted from an ultrafiltration unit.

In order to simulate an ultrafiltration process as the retentate becomes increasingly concentrated in both ion and surfactant, the permeate $Cu^{2+}$ concentration is shown in Table 1 and Fig. 4 as a function of retentate $Cu^{2+}$ concentration at a $SDS/Cu^{2+}$ mole ratio of 10. As the retentate concentration increases, the permeate becomes increasingly impure. As seen in Fig. 3, at the high surfactant concentrations present in Fig. 4, the surfactant concentration has little effect on permeate purity. Therefore, it is the increasing $Cu^{2+}$ concentration in the retentate in Fig. 4 that is primarily responsible for the decreasing permeate purity as the retentate becomes more concentrated. This also means that increasing surfactant concentration in the feed would not significantly improve the separation under these conditions. It is important to note that the rejection is approximately 99.5% over the entire path of the process shown in Fig. 4. Therefore, even at the end of the concentrating process, the permeate is still very pure under these conditions.

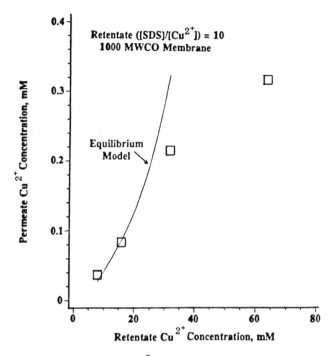

FIG. 4    Permeate $Cu^{2+}$ concentration holding the $SDS/Cu^{2+}$ ratio in the retentate constant.

## D.  Effect of Added Monovalent Salt

Many streams from which multivalent ions must be removed contain monovalent electrolytes as well.    The effect of added sodium chloride concentrations on permeate $Cu^{2+}$ concentrations is shown in Table 1 and Fig. 5.    The added salt can be very detrimental to the efficiency of metal removal using MEUF.    Addition of 0.1 M NaCl decreases the rejection from 99.53 to 84.75% and addition of 0.4 M NaCl results in further reduction to 61.62%.    Despite the fact that extremely high rejections are no longer attained, accept- able levels of separation can still exist at moderate levels of salinity.

In Sec. III.B, it was shown that increasing surfactant con- centration could decrease permeate metal concentrations in the absence of added salt.    However, under the conditions shown in

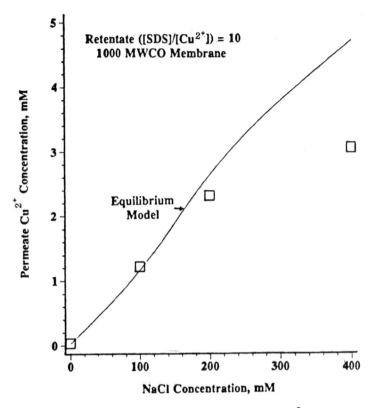

FIG. 5   Effect of added NaCl on permeate $Cu^{2+}$ concentration.

Fig. 3, the rejections were very high at all surfactant concentrations used (rejections increased from 98.54 to 99.77% as the
surfactant concentration was increased by a factor of 4). The
effect of increasing retentate surfactant concentration on permeate purities with 0.1 M added NaCl is shown in Table 1 and
Fig. 6; the data from Fig. 3 with no added salt are also shown
for comparison. In this case, as the retentate SDS concentration
is quadrupled from 40 to 160 mM, the rejection increases from
71.94 to 89.88%. As the surfactant concentration is quadrupled
again to 640 mM, the rejection increases to 96.16%. Therefore,
the deleterious effects of added monovalent electrolyte can be
substantially offset by the use of higher surfactant concentrations.

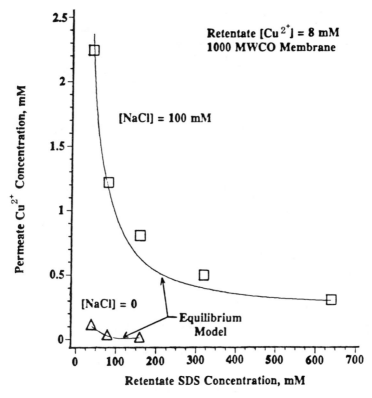

FIG. 6    Effect of retentate SDS concentration on permeate $Cu^{2+}$ concentration at different levels of added NaCl.


E.    Effect of Membrane Pore Size

Large-pore-size membranes are desirable in ultrafiltration because larger fluxes can be attained, resulting in a lower membrane area requirement and a lower capital cost and lower membrane replacement cost to treat a specific stream.    The effect of retentate composition on flux for a certain membrane will be discussed in Sec. III.J.    The effect of membrane pore size on $Cu^{2+}$ permeate concentration is shown in Table 1 and Fig. 7.    There is a decrease in the permeate metal concentration when the membrane molecular weight cutoff (MWCO) is increased from 1000 to 5000. A gel layer can form next to the membrane due to concentration

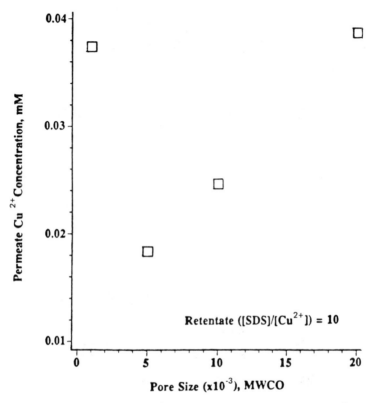

FIG. 7   Effect of membrane pore size on permeate $Cu^{2+}$ concentration.

polarization. This gel layer can contribute a presieving effect, resulting in improved permeate purities, as previously seen for organic removal using MEUF (3). Above a MWCO of 5000, very slight leakage of micelles occurs, but the effect is small (rejection decreases only from 99.77 to 99.51%). It may be concluded that MEUF can effectively block micelles and result in high rejections with membrane of MWCO of 20,000 or higher.

## F.   Effect of Multivalent Ion Type and Charge

The data discussed so far have been for divalent cationic copper. A comparision of results for MEUF of divalent cations $Cu^{2+}$, $Cd^{2+}$,

TABLE 2  Comparison of Removal of Various
Divalent Ions Using MEUF

| Metal | Permeate concentration (mM) | Rejection (%) |
|---|---|---|
| $Cd^{2+}$ | 0.084 | 98.9 |
| $Zn^{2+}$ | 0.058 | 99.2 |
| $Cu^{2+}$ | 0.037 | 99.5 |
| $Ca^{2+}$ | 0.034 | 99.5 |
| $CrO4^{2-}$ | 0.042 | 99.4 |

Temperature = 30°C; pressure drop = 414 kPa
(60 psig); retentate [metal] = 7.5 mM;
retentate [SDS] = 75 mM; 5000 MWCO
membrane.

$Zn^{2+}$, and $Ca^{2+}$ is shown in Table 2 (4). The permeate concentrations are nearly the same for each divalent cation under
the same conditions. Therefore, valence is the dominant characteristic determining the efficiency of removal of a multivalent ion
from water using MEUF. The rejections are in the order
$Ca^{2+} > Zn^{2+} > Cu^{2+} > Cd^{2+}$, although variations are small (rejections varied from 99.5 to 98.9% for these ions). The main
cause of these small differences is believed to be the complexation
of the cations with anions in solution. For example, Cd has a
much larger tendency to complex with chloride (co-ion present in
these experiments) than the other cations (5), leaving a smaller
concentration of $Cd^{2+}$ to bind to the micelles and thus be removed
from the permeate solution.

Also shown in Table 2 is the permeate concentration of the
divalent anion chromate using a cationic surfactant, cetylpyridinium
chloride (2). Approximately the same rejections are observed in
this case. Therefore, MEUF is effective in removal of either
multivalent cationic metals or anionic metallic complexes.

TABLE 3   Effect of Using an Anionic/Nonionic
Surfactant Mixture in Removal of Zinc in MEUF

| Mole fraction in micelle | | Rejection of $Zn^{2+}$ (%) | |
|---|---|---|---|
| Anionic surfactant | Nonionic surfactant | Actual | Predicted |
| 1.0 | 0 | 48.6 | 0 |
| 0.91 | 0.09 | 93.7 | 85 |
| 0.83 | 0.17 | 96.4 | 86 |
| 0.77 | 0.23 | 97.9 | 88 |

Temperature = 30°C; pressure drop = 414 kPa (60 psig);
retentate $[Zn^{2+}]$ = 0.13 mM, retentate [SDS] = 3.3 mM;
5000 MWCO membrane.

## G.  Use of Very Low Surfactant Concentrations

In many applications, the extremely large values of rejections
discussed so far are not required.  Often, a factor of 2 or 3
reduction in ion concentration is sufficient to meet environmental
emission standards.  If this is the case, it is important to exam-
ine methods of reducing the cost of a MEUF operation in attaining
this permeate concentration.  One obvious method is to reduce
the surfactant concentration in the feed.  Figures 3 and 6 show
that this will result in an increased multivalent ion concentration
in the permeate.  As the critical micelle concentration (CMC) is
approached, no micelles are present in the bulk solution and no
separation should occur in theory.

The first row in Table 3 corresponds to this situation.  However, 48.6% rejection was observed, not 0% as expected.  We
believe that this is a result of the formation of a gel layer next
to the membrane due to concentration polarization effects.  It
was shown in Chapter 1 that surfactant monomer can be rejected
to some extent in MEUF.  If this is the case, there may be an
accumulation of surfactant at the membrane surface.  The concentration of surfactant in this gel layer can exceed the CMC
and micelles can be present in this region, even when the bulk
retentate has no micelles present.  Concentration polarization
effects become more severe when transmembrane pressure

TABLE 4    Effect of Pressure on Zinc
Rejection at Low Surfactant Concentrations in MEUF

| Pressure (psig) | Rejection of $Zn^{2+}$ (%) |
|---|---|
| 20 | 59.6 |
| 40 | 71.0 |
| 60 | 83.3 |

Temperature = 30°C; retentate $[Zn^{2+}]$
= 0.14 mM; Retentate [SDS] = 5.4 mM;
5000 MWCO membrane.

differences are greater (3).  The effect of pressure drop on
permeate purities is shown in Table 4.  As the pressure drop
decreases and the gel layer has less tendency to form, the rejection of $Zn^{2+}$ decreases; this supports the gel layer explanation
of the observed effect.

From a practical viewpoint, this effect can permit a much
lower surfactant usage level for moderate reductions in ion concentration compared to that needed to attain 99.8% rejections.
This can make MEUF more attractive economically by reducing
raw material costs.

## H.  Use of Surfactant Mixtures

It is desirable to minimize the fraction of surfactant in monomeric
form because micelles are the effective separating agent in MEUF.
Despite the discovery of a separation at surfactant concentrations
below the CMC, mentioned in Sec. III.G, there is still a fundamental lower limit on surfactant observed, even with the gel
layer effect.  The monomer concentration in anionic surfactant
systems with micelles present (or the CMC of the system) can
be drastically reduced by the addition of small amounts of nonionic surfactant (6).

The effect of using mixtures of an anionic (SDS) and a nonionic surfactant (alkylphenol polyethoxylate with EO = 10) in the
removal of $Zn^{2+}$ is shown in Table 3 (7).  In these experiments,
the SDS concentration was held constant and the nonionic

surfactant mole fraction in the system varied. The monomeric
concentration of each surfactant was predicted by regular solution theory (6). Based on these values, the concentration of
micelles in the system was calculated and the predicted binding
from the model outlined in Sec. IV used to calculate the "predicted" rejection shown in Table 3. As discussed in Sec. III.G,
the anionic surfactant only experiment corresponded to no
micelles being present in the bulk solution, but the actual rejection was greater than 0%. The effect of adding small amounts
of nonionic surfactant is predicted to greatly increase rejections
(adding 9% nonionic surfactant should increase rejection from 0
to 85%). In the actual case, the nonionic surfactant did increase
rejection substantially (adding 9% nonionic surfactant increased
rejection from 48.6 to 93.7%). Therefore, the expected syner-
gism of using surfactant mixtures was observed, even though
actual separations were much better than predicted at all surfactant compositions. Rejections exceeding 90% can be attained
with extremely low total surfactant concentrations (<4 mM) by
utilizing mixtures.

## I.  Removal of Mixtures of Metals

Most practical applications in which metal must be removed from
water involve mixtures of metals. For example, in a metal-plating
industry wastewater, zinc and nickel might both be encountered.
For MEUF to be effective in these applications, the various metals
involved must be removed simultaneously.

The composition of the permeate in an experiment involving
the removal of a mixture of $Cu^{2+}$ and $Zn^{2+}$ with SDS as the surfactant is shown in Table 5 (4). Very effective removal of each
metal is observed under the conditions used. In fact, the total
metal concentration in the permeate is approximately equal to the
metal concentration in the permeate if a single metal were being
removed from a retentate with the same total metal concentration.
There is no deleterious effect of the metals being present as a
mixture. Those metals that are more efficiently removed as
single components have a lower permeate concentration when an
equimolar mixture of metals is treated as seen in Table 5. Similar results are observed for ternary mixtures of divalent cationic
metals (4).

## J.  Flux

Gel polarization effects can severly limit flux through the membrane during ultrafiltration (8). The flux during MEUF removal

TABLE 5    Removal of Metal Mixtures

| Metal | Permeate concentration (mM) | Rejection (%) |
|---|---|---|
| $Cu^{2+}$ (as single component) | 0.037 | 99.51 |
| $Zn^{2+}$ (as single component) | 0.058 | 99.23 |
| Total metal in mixture | 0.0439 | 99.41 |
| $Cu^{2+}$ in mixture | 0.0168 | 99.55 |
| $Zn^{2+}$ in mixture | 0.0271 | 99.28 |

Temperature = 30°C; pressure drop = 414 kPa (60 psig); in mixture, retentate $[Cu^{2+}]$ = 3.75 mM; retentate $[Zn^{2+}]$ = 3.75 mM; retentate $[SDS]$ = 75 mM; 5000 MWCO membrane.

of cadmium from water is shown in Fig. 8 as a function of retentate surfactant concentration. In solutions concentrated enough for concentration polarization to occur, the flux is a linear function of log(retentate concentration). Extrapolation of this relationship to the concentration at which the flux = 0 is the gel concentration of gel point. For this system, the gel concentration is 570 mM (this compares to 530 mM for CPC as discussed in Chapter 1). As long as the retentate surfactant concentration does not exceed approximately 200–300 mM, the flux is not substantially below that of pure water (particularly considering that the overall flux is an integral average between that for the feed surfactant concentration and the final concentration). Therefore, concentration polarization is not a severe problem in MEUF. This is the conclusion reached in Chapter 1 for organic removal by MEUF.

### K.  Simultaneous Removal of Multivalent Ions and Nonionic Organics

As discussed in Chapter 1, nonionic organic solute removal by MEUF depends on solubilization of the solute in the micelle interior. As discussed in this chapter, removal of multivalent ions by MEUF depends on electrostatic attraction between the ion and the charge micelles. In order to investigate the ability to remove *simultaneously* the metal divalent cations and organics,

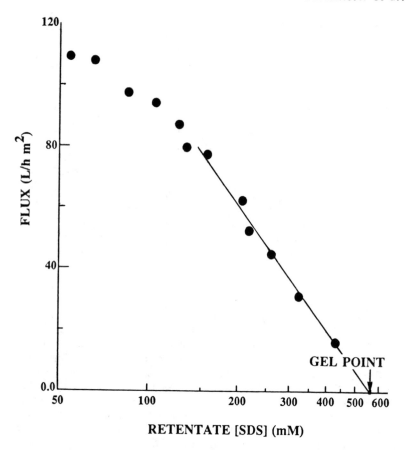

FIG. 8   Effect of retentate SDS concentration on flux.

removal of zinc and o-cresol by MEUF was studied individually and as a mixture.   Results are shown in Table 6 (9).

There is no significant effect of the presence of the metal on the removal of the organic and there is no significant effect of the presence of the organic on the removal of the metal.   Other systems studied show the same effects (9).   Therefore, in   mixtures of metals and nonionic organics, the correlations and models used in Chapter 1 can be used to estimate the removal efficiency of the organic and those discussed in this chapter can be used to estimate the efficiency of metal removal.

TABLE 6    Simultaneous Removal of Organics and Metal
Using MEUF

| Component | Permeate concentration (mM) | |
| | In single-component system | In mixture |
| --- | --- | --- |
| o-Cresol | 1.84 | 1.89 |
| $Zn^{2+}$ | 0.0056 | 0.0064 |

Temperature = 30°C; pressure drop = 414 kPa (60 psig);
retentate $[Zn^{2+}]$ = 1.5 mM; retentate [o-cresol] = 4.5 mM;
retentate [SDS] = 45 mM; 5000 MWCO membrane.

## IV.  THEORY

The concentration of a multivalent ion in the permeate stream is
related to that in the retentate under conditions assumed to exist
in MEUF separations.  In treating the separation of organic
solutes from aqueous streams by MEUF, it has been assumed that
the thermodynamic activity of each solute is the same in the
permeate and retentate solutions.  Chapter 1 described the
success with which this "equilibrium assumption" can be used to
predict MEUF results for a wide variety of solutes.

An analogous assumption makes it possible to predict the
results of MEUF separations of multivalent ions from aqueous
streams.  However, because the thermodynamic activities of salts
of multivalent ions involve both the ion and its counterion(s), it
is not possible simply to equate the concentration or activity of
an individual ion in the permeate to that in the retentate solution
under assumed equilibrium conditions.  But by equating the
thermodynamic activity of salts on the two sides of the membrane
in MEUF experiments, one can account for the retention of multivalent ions by surfactant micelles.  MEUF data for $Cu^{2+}$ (1) and
$CrO_4^{2-}$ (2) have been shown to be consistent with the equilibrium
assumption.

Christian et al. (2) adapted the spherical two-phase, poly-
electrolyte theory of Oosawa (10) to describe the binding of
counterions to ionic micelles.  In the Oosawa model, ions are considered as being either bound to the micelles (and hence unable
to pass through the membrane) or free in bulk aqueous phase

(and able to pass through the membrane as part of the permeate stream). A very simple theory based on this model is capable of correlating MEUF separation results for multivalent ions and their mixtures both in the absence and in the presence of added monovalent salts.

## A. Absence of Added Monovalent Electrolyte

The surfactant is normally produced as a salt consisting of the surfactant ion and a monovalent counterion. Therefore, in the simplest system, to remove a multivalent ion from water there will be two counterions present: the multivalent ion and the counterion from dissolution of the surfactant salt. Thus, if the divalent cation copper is being removed from solution using sodium dodecyl sulfate, $Cu^{2+}$ and $Na^+$ are the counterions in solution. The co-ion in this system is the ion associated with the multivalent counterion in the salt form. For example, if the copper is originally present as $CuCl_2$, $Cl^-$ is the co-ion. For convenience in the following expressions, we will assume that the surfactant is sodium dodecyl sulfate, the monovalent counterion is $Na^+$, the dodecyl sulfate is $DS^-$, the multivalent counterion is $Cu^{2+}$, and the monovalent co-ion is $Cl^-$. The equations can be used to describe other systems; this specific system is described by these equations for convenience. For this case, an equation can be written to describe the binding of each counterion to the micelle, Eq. (2) referring to the sodium and Eq. (3) to the copper

$$\ln \frac{[Na^+]_b}{[Na^+]_u} = \ln \frac{\Phi}{1 - \Phi} + P(1 - \beta)(1 - \Phi^{1/3}) \qquad (2)$$

$$\ln \frac{[Cu^{2+}]_b}{[Cu^{2+}]_u} = \ln \frac{\Phi}{1 - \Phi} + 2P(1 - \beta)(1 - \Phi^{1/3}) \qquad (3)$$

$$\beta \frac{[Na]_b + 2[Cu^{2+}]_b}{[DS^-]_{mic}} \qquad (4)$$

where the subscripts b and u denote the bound and unbound ions in the retentate, respectively, $\beta$ is the total fraction of the micellar charge which is neutralized by bound sodium and copper, $\Phi$ is the fraction of the total volume within which the bound counterions are located, $[DS^-]_{mic}$ is the dodecyl sulfate concentration in micellar form, and P is a dimensionless potential parameter. It should be noted that the $\beta$ used here is equal to $(1 - \beta)$ for the $\beta$ used by Oosawa (10) and Christian et al. (2); i.e., the $\beta$ used in that work was the total fraction of the micellar charge which is *not* neutralized.

The electroneutrality condition in the permeate and in the retentate leads to the following expressions:

$$2[Cu^{2+}]_{ret} + [Na^+]_{ret} = [DS^-]_{ret} + [Cl^-]_{ret} \tag{5}$$

$$2[Cu^{2+}]_{per} + [Na^+]_{per} = [DS^-]_{per} + [Cl^-]_{per} \tag{6}$$

where the subscripts "ret" and "per" denote the total concentrations of the designated ions in retentate and permeate, respectively.

Because the unassociated electrolyte is assumed to be in equilibrium across the membrane, the activity of any cation–anion pair must be equal in the permeate and in the retentate. In the absence of added electrolyte, the activity coefficients will be close to unity on both sides of the membrane. If it can be assumed that the activity coefficients are the same in the permeate and retentate, this equilibrium condition results in

$$[Cu^{2+}]_{ret}[Cl^-]^2_{ret} = [Cu^{2+}]_{per}[Cl^-]^2_{per} \tag{7}$$

$$[Cu^{2+}]_{ret}[DS^-]^2_{ret} = [Cu^{2+}]_{per}[DS^-]^2_{per} \tag{8}$$

$$[Na^+]_{ret}[Cl^-]_{ret} = [Na^+]_{per}[Cl^-]_{per} \tag{9}$$

The surfactant in the retentate can be present as monomer (unassociated molecules) or in micelles. In order to calculate the concentration of surfactant present as micelles (which is needed to calculate the value of $\Phi$), the following relationship is used (2):

$$\beta(\ln([Na^+]_u + 2[Cu^{2+}]_u)) + \ln[DS^-]_{mon} = (\ln(CMC))(1 + \beta) \tag{10}$$

where $[DS^-]_{mon}$ is the dodecyl sulfate present in monomeric form and CMC is the critical micelle concentration of the surfactant in the absence of any other added electrolytes. At the high surfactant concentrations normally used in MEUF, inaccuracies in Eq. (10) have little influence on the predicted permeate concentrations in MEUF.

## B. Presence of Added Monovalent Electrolyte

When added monovalent salt is present in the retentate in addition to the surfactant and the salt of the multivalent ion, the parameter P can be described by

$$P = \frac{P^0}{1 + \alpha[NaCl]^{1/2}} \tag{11}$$

where NaCl is used as the model added electrolyte, $P^0$ is the value of P in the absence of added monovalent electrolyte, and $\alpha$ is a constant. Equation (11) implicitly assumes that the co-ion from the added electrolyte is the same as that from the salt of the multivalent ion and that the counterion from the added electrolyte is the same as that from the salt of the surfactant (in this case sodium and chloride). Extension to cases where this is not truce is straightforward.

## C. Summary of Theory

If the value of P is known, e.g., calculated from Eq. (11), simultaneous solution of Eqs. (2) to (11) permits calculation of the unbound multivalent ion concentrations in the permeate, the variable of most practical concern in application of MEUF. Conversely, the value of P can be calculated from experimental MEUF results. Then the theory can be used to calculate permeate purities under different conditions of surfactant concentration, and multivalent ion concentration. The value of $\alpha$ in Eq. (11) can be obtained from permeate data in the presence of added salt. The solid lines in Figs. 2-6 are calculated from the theory using the values $P^0 = 55.1$ and $\alpha = 23.8$ for the SDS–$Cu^{2+}$ system.

The model fits the data reasonably well over a wide range of conditions (see Fig. 2) with the exception of high metal concentrations (Fig. 4) or high salinities (Fig. 5), where predicted permeate metal concentrations are higher than experimental values. The values of the constants $P^0$ and $\alpha$ will be different for different metal ions (4). Extension of this theory to describe removal of mixtures of metals is straightforward and the model describes those systems well also (4).

REFERENCES

1. J. F. Scamehorn, R. T. Ellington, S. D. Christian, B. W. Penney, R. O. Dunn, and S. N. Bhat, *AICHE Symp. Ser.*, *250*: 48 (1986).
2. S. D. Christian, S. N. Bhat, E. E. Tucker, J. F. Scamehorn, and D. A. El-Sayed, *AICHE J.*, *34*: 189 (1988).
3. R. O. Dunn, J. F. Scamehorn, and S. D. Christian, *Sep. Sci. Technol.*, *22*: 763 (1987).
4. D. A. El-Sayed, J. F. Scamehorn, and S. D. Christian, *Sep. Sci. Technol.* (in preparation).
5. R. M. Smith and A. E. Martell, *Critical Stability Constants*, Vols, 4 and 5, Plenum Press, New York, 1976.
6. J. F. Scamehorn, in *Phenomena in Mixed Surfactant Systems* (J. F. Scamehorn, ed.), ACS Symp. Ser., Vol. 311, 1986, p. 1.
7. L. W. Brant, J. F. Scamehorn, and S. D. Christian, *J. Colloid Interf. Sci.* (in preparation).
8. M. C. Porter, in *Handbook of Separation Techniques for Chemical Engineers*, McGraw-Hill, New York, 1979, Sec. 2.1.
9. D. O. Dunn, J. F. Scamehorn, and S. D. Christian, *Colloid Surf.* (submitted for publication).
10. F. Oosawa, *Polyelectrolytes*, Marcel Dekker, New York, 1971.

# II
# SEPARATIONS BASED ON EXTRACTION

# 3

# Reversed Micellar Extraction of Proteins

**T. ALAN HATTON**   Department of Chemical Engineering,
Massachusetts Institute of Technology, Cambridge, Massachusetts

The author gratefully acknowledges the support of the NSF
Biotechnology Process Engineering Center at M.I.T. during the
preparation of this manuscript.

SYNOPSIS

Reversed micelle-containing organic solvents can be effective
extractants for the recovery, purification, and concentration of
proteins using traditional liquid-liquid extraction techniques.
Factors governing the solubilization of proteins by these novel
solvent systems include the electrostatic interactions between the
proteins and the charged surfactant headgroups, and the aggregative
properties of the surfactants in solution. Electrostatic interactions
can be mediated by varying pH, ionic strength, and salt and
buffer type. Solvent structure and surfactant type are important
in determining the cooperative formation of the protein-micelle
complex. Enhanced partitioning of selected proteins is obtained
through the incorporation of biospecific affinity surfactants in the
surfactant-solvent mixtures. This partitioning behavior has been
described effectively in terms of simple phenomenological thermodynamic models. Applications of reversed micelles in the recovery
of both intracellular and extracellular proteins have provided
promising results. Efficient contacting of the two phases can be
achieved using conventional mixer-settler systems, or the newer
hollow-fiber membrane extractors.

## I. INTRODUCTION

In recent years, increased attention has been given to the development of efficient methods for the separation, concentration, and
purification of proteins and other bioproducts from fermentation
and cell culture media. This is a result of the rapid progress made
in recombinant DNA techniques and genetic engineering, advances
that promise many new and exciting applications for biotechnology.
The traditional methods for the recovery of biomolecules from
complex mixtures, such as chromatography and electrophoresis,
were developed for small-scale analytical or preparative applications.
While these methods can be used for larger-scale separations, in
many cases alternative isolation and purification procedures may
be desirable. This is particularly so if these bench-scale methods
suffer from a loss of resolution on scale-up or do not offer the
advantages of economy of scale observed in the traditional chemical
process industries.

Liquid-liquid extraction is a technique that, although used
extensively in the antibiotics industry, has not found significant
application in other sectors of the biotechnology community. One
reason for this is the lack of suitable solvents having the desired

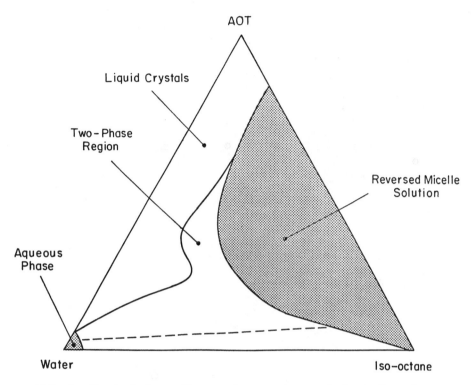

FIG. 1    Typical phase diagram for a water-surfactant-oil system.

selectivity and capacity for the products of interest, and which
are gentle on sensitive tertiary structures that can be permantly
disrupted if exposed to hostile environments.   Moreover, the
multi-ionic character of proteins and other bioproducts precludes
their direct extraction by the organic solvents traditionally used
in solvent extraction operations.   The addition of simple ion-
pairing carriers to these solvents will not work in most cases
(except at extremes of pH) since charges of both polarities are
generally present on biomolecules and they cannot all be ion-
paired simultaneously.

A new class of solvents is beginning to emerge that appears
to offer many advantages for the selective recovery, concentration,
and purification of proteins and other biomaterials using liquid
extraction technology.   These solvents rely on the unique

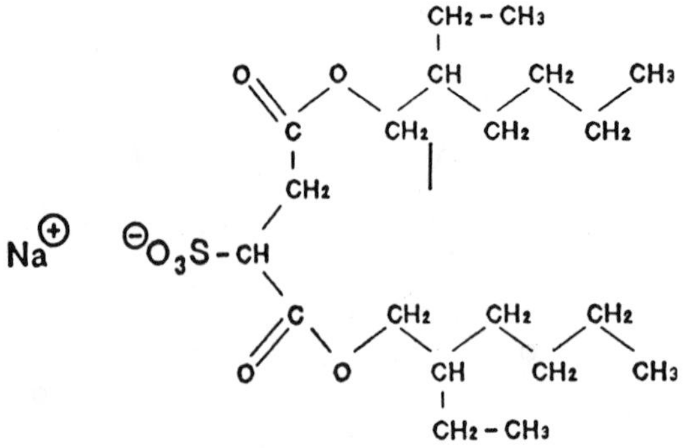

(a)

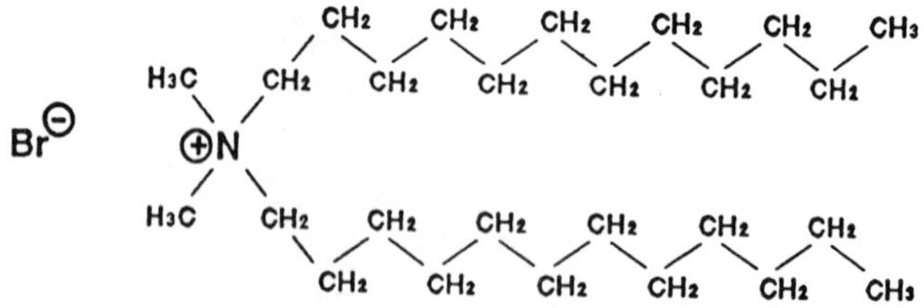

(b)

FIG. 2  Surfactants used in the reversed micellar extraction of proteins (a) sodium di-2-ethylhexyl sulfosuccinate (AOT), (b) didodecyldimethylammonium bromide (DDAB), and (c) trioctylmethylammonium chloride (TOMAC).

solubilizing properties of many surfactants that tend to aggregate in organic solvents to form reversed micelles.  The aggregates consist of a polar core of water and solubilized species stabilized by a surfactant shell layer.  Depending on the relative concentrations of surfactant, water, and oil, these mixtures can show a rich variety of structures, ranging from normal micellar formation, through the formation of reversed micelles and hexagonal rod-shaped structures, to lamellar or liquid crystalline phases.  A simplified depiction of this rich phase structure is given in Fig. 1. The region of current interest from the protein recovery standpoint is the two-phase region at the bottom of the phase diagram in which any mixture of the components will result in two bulk equilibrium phases, one being the aqueous feed solution containing

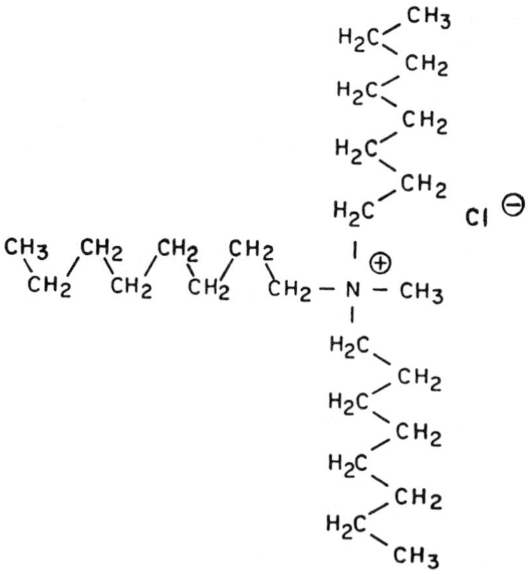

(c)

FIG. 2    (Continued)

some solvent and surfactant in dilute quantities, and the other being the reversed micellar phase which serves as the extractant in the protein product recovery operation.  These two resulting phases are indicated by the tieline shown on this figure; any mixture falling on the tieline will phase-separate to form two phases of the compositions indicated by the intersections of the tieline with the two-phase envelopes.  All studies to date on protein extraction using reversed micelles have concentrated on this region of the phase diagram, although there may be some incentive to encroach on the other phase structure regimes. Surfactants typically used in extraction operations are shown in Fig. 2.

A schematic representation of the reversed micellar extraction of proteins is shown in Fig. 3.  The effective utilization of these extractant solutions for large-scale product recovery will depend on the partitioning behavior of the different componenets in the feed stream between the two phases, and on process considerations such as the relative ease with which the phases can be

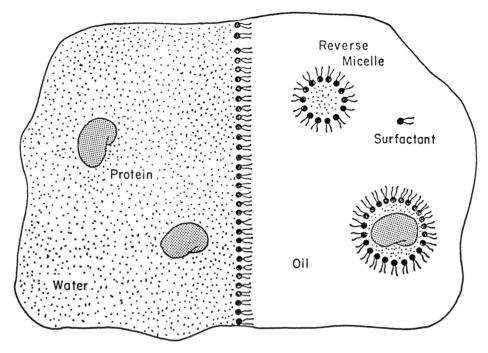

FIG. 3   A schematic representation of the reversed micellar
extraction of proteins.

contacted and subsequently separated.  It is the purpose of this
chapter to address some of these issues and to place them within
the framework of current knowledge on these topics.

## II.  PROTEIN SOLUBILIZATION

Luisi and coworkers (1–3) were the first to recognize the potential for separating and purifying proteins based on their ability
to transfer selectively from an aqueous solution to a reversed
micelle-containing organic phase and to be subsequently recovered
in a second aqueous phase.  It is only recently, however, that
this process has been looked at systematically by Dekker et al.
(4–8) and Hatton and coworkers (9–19) as a separation technique
for the large-scale recovery of proteins from aqueous media and

that a clearer picture has begun to emerge of the factors affecting the transfer of proteins in different micellar systems. The current state of understanding of the solubilization phenomena is reviewed in this section.

## A. Factors Affecting Protein Solubilization

The partitioning of proteins between a bulk aqueous phase and a reversed micellar solution will depend on the aqueous phase conditions of pH, ionic strength, and salt type, all of which affect the physicochemical state of the protein and its interaction with the aqueous solvent and the surfactant headgroups. In addition to these factors, solvent structure and surfactant type, which play a significant role in determining the aggregative properties of the surfactants in solution, will influence the protein-partitioning behavior inasmuch as they will affect the cooperative formation of the protein-micelle complex. While progress has been made in elucidating some of the effects these various parameters have on protein-partitioning behavior, the data are still too fragmented to enable definitive conclusions to be drawn.

## 1.  Effect of pH

The pH of the solution should affect the solubilization character- istics of a protein primarily in the way in which it modifies the charge distribution over the protein surface. Secondary factors, such as changes in protein conformation as a result of pH swings, may also play a role in this regard. At pH values below its isoelectric point (pI), or point of zero net charge, the protein will take on a net positive charge, while above its pI the protein will be negatively charged. Thus, if electrostatic interactions are the dominant factor in the solubilization process, solubilization should be possible with anionic surfactants only at pH values less than the pI of the protein, where electrostatic attractions between the protein and surfactant headgroups are favorable. At pH values above the pI, electrostatic repulsions would inhibit the protein solubilization. The reverse trends would be anticipated in the case of cationic surfactants.

In early work using the anionic surfactant AOT (Aerosol-OT; sodium di-2-ethylhexylsulfosuccinate), Luisi and coworkers (2) did not observe these trends with ribonuclease A. They found the transfer to be more efficient in the alkaline pH range and therefore concluded that electrostatic interactions are not impor- tant in the overall solubilization process. Goklen and Hatton (12)

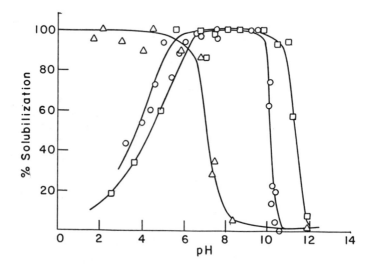

FIG. 4    The effect of pH on the solubilization of (□) lysozyme, (○) cytochrome c, and (△) ribonuclease A in AOT-isooctane solutions. Salt concentration: 0.1 M KCl. (Reprinted from Ref. 12, with permission.)

found the opposite to be true for the three low molecular weight protines ribonuclease A, cytochrome c, and lysozyme. Indeed, as shown in Fig. 4, with a 50 mM AOT solution, there was a very strong transition in solubilization behavior at the pI of each protein, which demarcated the region of no solubilization (pH > pI) from that of almost quantitative transfer to the organic solution (pH < pI). This observation is in accord with the postulated solubilization mechanism based on electrostatic interactions.

For larger proteins, at the same levels of surfactant loading, the pH had to be reduced to values significantly below the pI for there to be any appreciable solubilization. However, the earlier behavior, in which the pI marked the point of transition between the solubilization regimes, could be recovered by increasing the surfactant concentration (11,13,15). As discussed later, this effect can be interpreted in terms of a mass action kinetics model in which the proteins and micelles are in equilibrium with a protein-micelle complex.

Studies using cationic surfactants have focused primarily on the quaternary ammonium salt trioctyl methyl ammonium chloride (TOMAC). Luisi et al. (1) extracted trypsin, α-chymotrypsin, and pepsin using TOMAC in cyclohexane. There was little correlation between the extraction behavior and the protein pI in these studies. The sensitivity of the extraction process to the structural qualities of the protein being extracted was emphasized by the lack of solubilization of α-chymotrypsinogen even under conditions for which significant transfer of α-chymotrypsin was obtained. A number of other proteins could not be solubilized using TOMAC, or were only slightly solubilized, including lysozyme, peroxidase, myoglobin, and yeast and horse liver alcohol dehydrogenases. These authors also noted a decrease in the solubilization in the presence of buffers. As expected, solvent selection and temperature changes were also observed to affect the overall transfer of proteins to the reversed micellar phase.

A systematic study was recently undertaken by Dekker and coworkers (4–7), who concentrated on α-amylase extraction into TOMAC/isooctane solutions. They have investigated the effects of pH (4,5), salt concentration (7), and the addition of a nonionic cosurfactant (6) on the solubilization profile for the enzyme, with the results as shown in Fig. 5. In the first two cases, a narrow solubilization peak was observed in the alkaline pH range, with the peak shifted to an even higher pH range as the salt concentration was increased. The addition of the nonionic surfactant nonylphenolpentaoxylate resulted in a more efficient extraction of the enzyme into the organic phase, and in a broadening of the pH range over which solubilization was possible. In addition, a slow inactivation of the enzyme by TOMAC was observed to be significantly reduced on this addition of the nonionic surfactant.

Goklen (11) used the surfactant didodecyldimethylammonium bromide (DDAB) in trichlorethylene to solubilize the three enzymes lysozyme, ribonuclease A, and carbonic anhydrase. These proteins were selected because of their widely varying isoelectric points (11.0, 7.8, and 5.5, respectively). The results are shown in Fig. 6, where it is evident that with this cationic surfactant there appears to be no discernible selectivity in the extraction, based on the pI values of the proteins. The extraction curves for all three proteins were very similar, exhibiting significant transfer to the organic phase only at the high pHs, and no discrimination between proteins in this extraction. Under these conditions, all amino surface residues can be anticipated to be

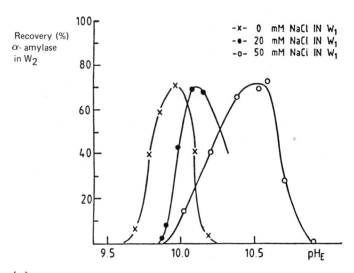

(a)

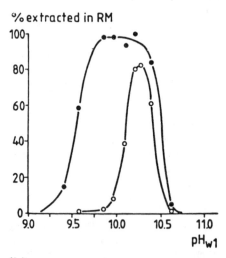

(b)

FIG. 5 The effect of pH on the solubilization of α-amylase in a TOMAC/isooctane solution (a) at different salt concentrations and (b) with (●) and without (○) the addition of the nonionic surfactant nonylphenolpentaoxylate. (Reprinted from Refs. 6 and 7 with permission.)

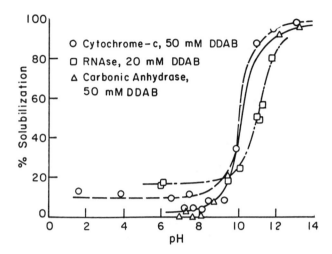

FIG. 6    The effect of pH on the solubilization of cytochrome c, ribonuclease A, and carbonic anhydrase by the cationic surfactant DDAB in trichlorethylene.   (Reprinted from Ref. 11.)

deprotonated, leaving only negatively charged surface residues available for participation in electrostatic interactions with the cationic surfactants.   This suggests that the solubilizing mechanism may not be entirely via reversed micelles, but that simple ion pairing of the surfactants with the anionic surface residues may be the dominant means for extraction.

2.   Ionic Strength Effects

The ionic strength of the aqueous solution in contact with the reversed micellar phase affects the protein-partitioning behavior in a number of ways.   The first is through the mediation of the electrostatic interactions between the protein surface and the surfactant headgroups via modification of the properties of the electrical double layers adjacent to both the charged inner micellar wall and the protein surface.   An increase in ionic strength compresses the range over which electrostatic interactions can overcome the thermal motion of the solute molecules, and thus decreases the protein-surfactant interactions, inhibiting the solubilization  of the protein.   This electrostatic screening effect is also responsible for reducing the surfactant headgroup

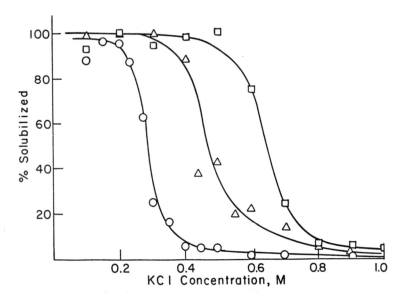

FIG. 7    The effect of [KC1] on the solubilization of (□) lysozyme,
(○) cytochrome c, and (△) ribonuclease A in AOT/isooctane
solutions.    (Reprinted from Ref. 12, with permission.)

repulsions, leading to the formation of smaller reversed micelles.
This can lead to a decrease in solubilization capacity through a
size exclusion effect.

A third effect of ionic strength is to salt out the protein from
the micellar phase because of the increased propensity of the
ionic species to migrate to the micellar water pools and to displace
the protein.    Finally, specific and nonspecific salt interactions
with the protein or surfactant can modify the solubilization behavior, and these effects will be more pronounced the higher the
salt strength.

Strong ionic strength effects have been observed in protein
solubilization experiments by all three groups working on the
phase transfer problem.    The effects of [KC1] on the solubilization
of the three proteins ribonuclease A, cytochrome c, and lysozyme
are shown in Fig. 7 (12).    The solubilization behavior is sensitive
to ionic strength over a narrow range, beyond which the protein
is either completely solubilized or its solubilization is almost
totally inhibited.    This range is dependent on the protein species

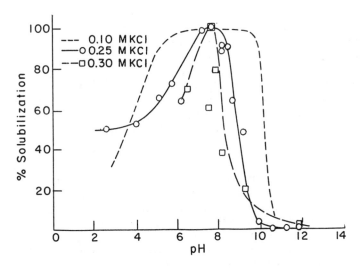

FIG. 8   The effect of [KC1] on the pH solubilization peaks for cytochrome c. (Reprinted from Ref. 11.)

and does not correlate with the protein net charge in the aqueous solution, as reflected in the departure of the pH used in the experiments from the natural pI of the protein.  Thus the ribonuclease A solubilization curve is intermediate between those of the two other proteins used in this work.  While no definite conclusions can be drawn as to the reasons for this behavior, it can be speculated that it relates to the protein surface characteristics, in particular to the distribution of charged residues and hydrophobic regions over the protein surface.  These results point to the sensitivity of the extraction process to the structural attributes of the targeted proteins, and indicate the potential for tailoring the feed solution to maximize the selectivity of the extraction.

A more detailed study of the combined effects of pH and ionic strength on protein solubilization demonstrated that the net effect of the increased salt concentration was to increase the pH excursion from the pI required for significant protein solubilization to occur.  The results for cytochrome c are shown in Fig. 8, where it is evident that the pH solubilization region is narrower at higher ionic strengths than at the lower salt concentrations.

With the anionic surfactant AOT it has been noted that as salt concentration increases, it is necessary to use lower pH values to effect significant solubilization. Dekker et al. (7) found the opposite to be true with the cationic surfactant TOMAC, the solubilization peak shifting to higher pH values as the NaCl concentration was increased. This observation is consistent with the notion that electrostatic interactions are dominant in these systems, and increased ionic strengths affect the solubilization through any or all of the mechanisms cited above.

### 3. Effect of Salt Type

Salt type can play an important role in determining the solubilization characteristics of different proteins. This was clearly demonstrated in the works of Luisi and coworkers (2,3,20). They found that salt and buffer type were important factors in determining transfer across an organic solvent bridge between two aqueous solutions of different pH and ionic strengths. The results from their most recent study are reported in Table 1, which clearly indicates that different salts have decidedly different effects on transfer efficiency (20).

Goklen (11) also observed significant effects of salt type. With $CaCl_2$ as the salt species, the pH range over which significant solubilization of cytochrome c occurred increased, probably due to $Ca^{2+}$ binding to the protein or to the surfactant. This would modify the electrostatic interactions between the protein and surfactant. The choice of buffer can also affect the solubilization characteristics of certain proteins. Use of a potassium phosphate buffer to adjust pH resulted in a shift of the pH solubilization curve to higher pH values relative to those required for solubilization when pH was adjusted by simple acid–base addition. This again points to the importance of all components in the feed solution in determining the effectiveness and selectivity of the extraction process.

### 4. Surfactant Concentration Effects

An enhanced capacity of the reversed micellar phase for proteins can be achieved by increasing the surfactant concentration. This was clearly observed by Goklen and Hatton (11,13,15), who noted that the pH solubilization peak broadened as the surfactant concentration was increased. In other studies, Dekker et al. (6) found a marked improvement in the pH range over which α-amylase could be solubilized by the cationic surfactant TOMAC by incorporating a nonionic surfactant within the solvent formulation.

TABLE 1  Phase Transfer of Proteins from Aqueous into Micellar Solutions[a]

| Salt | pH | RNase-A | Lysozyme | Trypsin | Water |
|------|-----|---------|----------|---------|-------|
| | | | Percentage of Transfer | | |
| 1 M CaCl$_2$ | — | 15.7 ± 5 | 100.9 ± 5 | 31.3 ± 9 | 0.87 ± 0.12 |
| 0.1 M CaCl$_2$ | 10 | 7.6 ± 2 | 98.5 | 27.0 ± 1 | 1.23 ± 0.10 |
| 0.1 M CaCl$_2$ | 5 | 96.0 ± 3 | 103.0 ± 6 | 59.1 ± 6 | 1.29 ± 0.09 |
| 1 M KCl | 5 | 4.0 ± 1 | 11.5 ± 7 | 14.4 ± 2 | 0.45 ± 0.07 |
| 0.1 M MgCl$_2$ | — | 86.6 ± 5 | 9.3 ± 1 | 21.4 | 3.42 ± 0.04 |

[a]The initial concentration of the protein in the water phase is about 15 μM. The overall concentration of the protein in the micellar solution is 15 μM for 100% transfer. The results are the average of several determinations.
Source: Ref. 20.

The results are shown in Fig. 5. This enhanced solubility range
is probably a consequence of the structural changes in the reversed micelle that are known to occur with mixed surfactant
systems such that larger reversed micelles able to accommodate
the hosted protein are formed.

Woll and Hatton (14,17) investigated the equilibrium solubilization characteristics for the two proteins ribonuclease A and
concanavalin A at different AOT concentrations over a range of
pH values below the pI values of the proteins. The increase
in solubilization with increasing surfactant concentration was
described in terms of a phenomenological thermodynamic model
based on mass action kinetics, where a protein P interacts with
n empty micelles M to form a filled protein-micelle complex PM
according to the pseudoreaction

$$P + nM \rightleftharpoons PM \tag{1}$$

The free energy change on solubilization is related to the
equilibrium constant for this pseudoreaction via

$$\frac{-\Delta G}{RT} = \frac{-(\Delta G^{\circ} + zF\Delta\psi)}{RT} = \ln\frac{[PM]}{[P][M]^{n}} \tag{2}$$

where z is the protein net charge, F is Faraday's constant, and
$\Delta\psi$ is the interphase distribution potential resulting from the surface excess of the charged surfactants at the bulk interface. The
micelle concentration is related to surfactant concentration [S] by

$$[M] = \frac{[S]}{N_{ag}} \tag{3}$$

where the aggregation number $N_{ag}$ is assumed to be independent
of [S]. The protein charge z will depend on the solution pH; to
a first approximation this dependency can be assumed to be
linear. Also, the size of the protein-micelle complex, of which n
is a direct measure, will depend on the strength of the electrostatic
interactions between the protein and the charged surfactant wall.
The greater this interaction, the smaller will be the protein-micelle
complex. Thus, n can be anticipated to decrease with increasing
z, or decreasing pH. Again a linear dependency on pH is assumed.
Based on these arguments, it was possible to develop an expression
in terms of pH and [S] for the partition coefficient

$$K = \frac{[PM]}{[P]} \tag{4}$$

The resulting equation is

$$\ln(K) = A + B(pH) + (C + D(pH)) \ln[S] \tag{5}$$

where A,B,C, and D depend on the interphase distribution potential, the slope of the protein titration curve, and the standard state free change on interphase transfer.

This equation was found to be particularly useful in correlating the solubilization data for ribonuclease A and concanavalin A over the pH ranges where the partition coefficient varied between 0.2 and 8, as shown in Fig. 9. Beyond these ranges, both experimental precision and deviations from the assumed linearity of charge z and n with pH yielded deviations from the correlation. Values for the constants for the two proteins are given in Table 2.

These results illustrate how a simple modeling approach can be used to good effect in the correlation and interpretation of solubilization data, and are the first to attempt to quantify solubilization behavior in rational thermodynamic terms. It remains to be seen how this approach can be extrapolated and extended to other systems.

## B.  Affinity Partitioning

Significant enhancements in the selectivity of protein reversed micellar extractions can be achieved by exploiting the molecular recognition capabilities of proteins for very specific ligand molecules of the type used, e.g., in affinity chromatography. The ligand can be covalently bound to a long alkyl chain such that the resulting amphipathic molecule will be positioned within the micellar wall, with the hydrophilic affinity group protruding into the polar core. Selective binding of this ligand (which for enzymes is frequently a substrate or inhibitor) to the traget protein in the mixture results in its preferential extraction into the reversed micelle core. The alkyl tail group favors the anchoring of the affinity surfactant within the extractant phase and deters it from being stripped into the aqueous phase. The basic concepts are illustrated schematically in Fig. 10.

The concentration of ligand surfactant molecules required in the surfactant-solvent mixture will be dictated in part by the strength of the interactions between the targeted protein and the

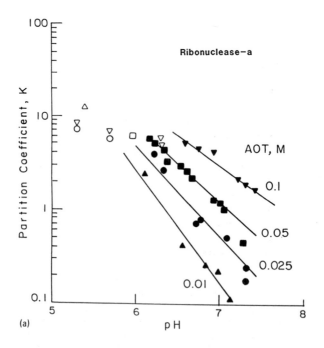

FIG. 9  The correlation of protein partition coefficients with
pH and AOT concentration (a) ribonuclease A, (b) concanavalin A.
(Reprinted from Ref. 17.)

ligand, i.e., on the binding or equilibrium constant for this
interaction.  In general, it is anticipated that these affinity
surfactants will be required in concentrations low relative to that
of the total surfactant present.

The effectiveness of this affinity partitioning of proteins in
reversed micelles has been demonstrated with the jack bean lectin,
concanavalin A, a carbohydrate-binding protein (14,15,18).  The
ligand used was the commercially available biological surfactant
octyl-β-D-glucopyranoside, which is insoluble in isooctane but
is moderately soluble in an AOT solution, where it presumably
behaves as a cosurfactant.

Typical affinity partitioning results are shown in Fig. 11.  It
is evident that the degree of protein transfer is dictated strongly

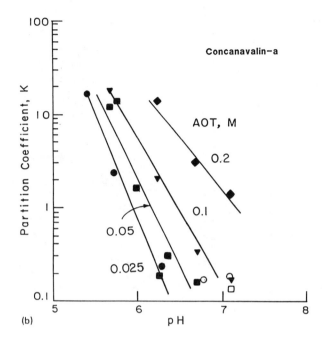

FIG. 9 (Continued)

TABLE 2 Correlation of Partition Coefficients
with pH and Surfactant Concentration:
$\ln(K) = A + B(pH) + (C + D(pH)) \ln[AOT]$

|   | Ribonuclease A | Concanavalin A |
|---|---|---|
| A | 4.43 | 11.42 |
| B | −0.061 | − 1.054 |
| C | −2.84 | − 5.50 |
| D | 0.585 | 1.16 |

*Note*: [AOT] is in molar units.

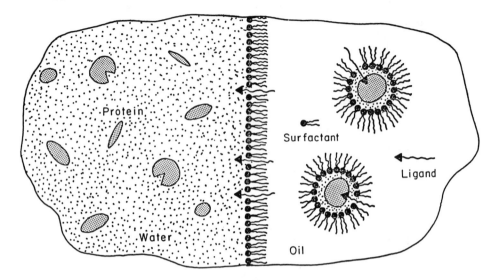

FIG. 10  The concept of affinity partitioning in reversed micelles.

by the biosurfactant concentration. This was not an artifact resulting from the formation of mixed micelles having different solubilization characteristics from the original AOT micelles, since the
presence of the biosurfactant did not significantly affect the partitioning of ribonuclease A under the same conditions. Similarly,
when the concanavalin A solution was swamped with glucose such
that all binding sites were occupied, the biosurfactant did not
affect the transfer of this protein, again indicating that the site-
specific interactions of the ligand with the protein are necessary
for enhancement of the extraction efficiency.

The dependency of the protein partition coefficient on the ligand
concentration [L] can be deduced through simple thermodynamic
arguments following the approach of Flanagan and Barondes (21).
For a single binding site, the affinity partitioning process can be
envisaged as shown in Fig. 12. The protein-ligand complex in
the bottom (aqueous) phase dissociates to give the individual free
ligand and protein species. These species partition independently
to the micellar phase, where they are able to recombine to form
the protein-ligand complex in this phase. If the dissociation

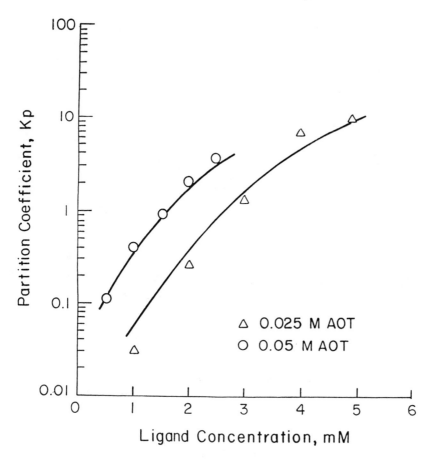

**FIG. 11** The effect of ligand concentration on the partition efficient for concanavalin A in AOT/isooctane solutions at pH 7.2. (Reprinted from Ref. 18.)

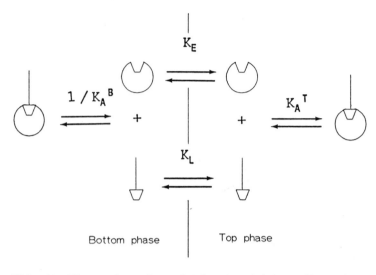

Bottom phase          Top phase

FIG. 12 Thermodynamic model for determining effect of ligand concentration on affinity partitioning of proteins.

constants in the two phases are $K_{DW}$ (water) and $K_{DM}$ (micellar), respectively, and the protein and ligand have respective partition coefficients $K_O$ and $K_L$, then it is readily shown that

$$K = \frac{([P] + [PL])_M}{([P] + [PL])_W} = K_o \left[ \frac{1 + [L]/K_{DM}}{1 + [L]/K_{DW}K_L} \right] \qquad (6)$$

If there are n identical binding sites per molecule, then it can be shown that the bracketed term must be raised to the nth power.

Under the conditions used in their experiments, Woll et al. (14,18) observed that concanavalin A was present only as a dimer having two binding sites per molecule. Thus, the total protein partitioning should depend on ligand concentration according to

$$K = K_o \left[ \frac{1 + [L]/K_{DM}}{1 + [L]/K_{DW}K_L} \right]^2 \qquad (7)$$

The protein partition coefficient $K_0$ has been shown to depend on pH and surfactant concentration via the approximate expression [see Eq. (5)]:

$$K_0 = K_0' (pH) [S]^{\alpha(pH)} \tag{8}$$

The effective surfactant concentration [S] is assumed to depend on the component surfactant concentrations through the linear additive relation

$$[S] = [AOT] + \beta[L] \tag{9}$$

The quantity $\beta$ is the molar AOT-equivalent solubilizing power of the ligand and is anticipated to be a direct function of the ratio of the surfactant head areas for the two surfactant types.

With these assumptions, and assuming a strong partitioning of the ligand to the micellar phase ($K_L$ very large), the protein partition coefficient should depend on ligand concentration as

$$K = K_0' ([AOT] + \beta[L])^{\alpha} (1 + [L]/K_{DM})^2 \tag{10}$$

At a pH of 7.2 $K_0'$ and $\alpha$ were estimated from Eq. (5) to be 46.5 and 2.84, respectively, where the surfactant concentrations are in molar units. A nonlinear regression analysis of the affinity partitioning data yielded values for $\beta$ and $K_{DM}$ of 8.8 and $2.7 \times 10^{-4}$ M, respectively. The good fit of the data to the model is illustrated in Fig. 11.

These early results are a clear indication of the promise that affinity partitioning holds for providing added flexibility in the reversed micellar extraction of bioproducts. They also point to the effectiveness of simple thermodynamic arguments in quantifying and interpreting complex solubilization phenomena. The challenge now is in the synthesis of suitable affinity surfactants for other protein systems, and in the development of more rigorous modeling approaches to correlate and predict solubilization behavior.


## III. PRODUCT RECOVERY AND ACTIVITY

The effectiveness of reversed micellar extraction for the large-scale recovery of proteins will depend on the ease with which the protein can be stripped from the loaded organic phase, and on the degree to which enzymatic activity or biological function is retained by the recovered product. There has been no systematic study of this topic to date, and in many cases it is found that different proteins respond differently to variations in stripping solution properties such as pH and ionic strength. Indeed, many proteins are not able to be recovered from the organic phase by simple manipulation of these solution properties and other measures need to be undertaken to effect product recovery, as is discussed below. For proteins that can be recovered directly, it is usually sufficient to adjust the pH and/or ionic strength of the stripping solution to conditions unfavorable for protein solubilization to strip the protein from the reversed micellar phase. A number of proteins fall in this category; in particular cytochrome c, α-chymotrypsin, and α-amylase, among others, have been recovered efficiently using this approach. In the case of lysozyme, however, we have found it necessary to go to extreme values of pH and ionic strength to obtain complete recovery of the protein product. Salt type can also have an effect on the protein efficiency.

In the work of Woll et al. (14,17,18), the proteins ribonuclease A and concanavalin A were not able to be recovered at all using pH and ionic strength variations, indicating a very strong protein-surfactant interaction in these cases. It was observed, however, that by adding a small quantity of a more polar, water-immiscible solvent to the reversed micellar solution, the reversed micelles could be disrupted and the protein completely recovered (14). In particular, using 10–20% by volume of ethyl acetate mixed with the isooctane/AOT solution, complete recovery of both concanavalin A and ribonuclease A could be obtained with no apparent loss in activity. One drawback associated with this approach is that the solvent cannot be reused for the extraction step, since its solubilizing power will have been reduced drastically through the addition of the polar organic. Solvent regeneration should be possible, however, using distillation to separate the ethyl acetate from the isooctane/AOT solution.

The retention of biological activity during the extraction/stripping cycle is an important consideration in the large-scale implementation of reversed micellar extraction processes. While in some cases activity loss can be severe, it is more often observed that there is a significant degree of activity retention, as reported by all three groups working in this area.

## IV.  PROTEIN SEPARATIONS USING REVERSED MICELLES

The ability to control the solubilization of different proteins through the manipulation of process parameters including pH, ionic strength, and surfactant type and concentration indicates the potential for controlling the selectivity of protein extraction from solution mixtures, such that the desired protein can be obtained in purer and possibly more concentrated form than in the original feed mixture.  There have been relatively few studies on the separation of proteins, the majority of the work having been done at MIT, with both synthetic protein mixtures and real fermentation media.  In this section, the results from these studies are summarized.

### A.  Synthetic Mixtures

The ease with which certain protein mixtures can be separated using reversed micellar extraction was clearly demonstrated by Goklen and Hatton (10–12), who investigated a series of binary protein mixtures and one ternary mixture.  In the latter case, judicious selection of the feed conditions, based on the results of Figs. 4 and 7, enabled the quantitative extraction of cytochrome c and lysozyme from a ternary mixture of these proteins with ribonuclease A.  The ribonuclease A was retained in the aqueous raffinate phase.  Two subsequent stripping steps, in which the loaded micellar organic solution was contacted sequentially with aqueous stripping solutions of different pH and ionic strength conditions, resulted in first the complete recovery of cytochrome c and then the recovery of lysozyme, in separate aqueous solutions.  The almost complete resolution of the ternary mixture is evident from the reversed phase PHLC analyses of the feed solution, and of the final three aqueous solutions, as shown in Fig. 13.

Woll and Hatton (14,17) investigated the separation of mixtures of ribonuclease A and concanavalin A showing that, in this case at least, the system behaved ideally, and that there were no interactions between the proteins.  In extending their work to the affinity partitioning scheme, they furthermore showed that the selectivity for one protein over another via protein-specific interactions with appropriate affinity surfactants could be predicted based on the single-protein extraction studies (18).

These studies were the first to demonstrate the feasibility of using reversed micelles to separate proteins in aqueous mixtures,

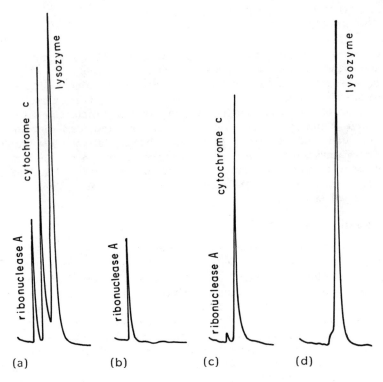

FIG. 13   HPLC analysis of (a) feed, (b) raffinate, pH 9
[KCl] = 0.1 M, (c) first stripping solution, [KCl] = 0.5 M, and
(d) second stripping solution, pH 11.5, [KCl] = 2.0 M in the
separation of a ternary mixture of the proteins ribonuclease A,
cytochrome c, and lysozyme.   (Reprinted from Ref. 12 with
permission.)

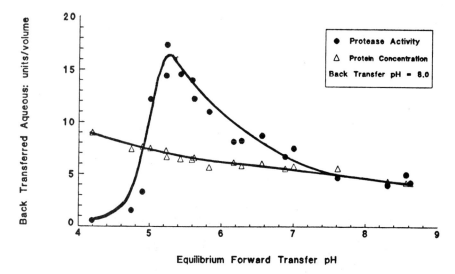

FIG. 14   The effect of feed pH on the extraction of protein in the recovery of an alkaline protease from a fermentation medium: ($\triangle$) total protein mass extracted, ($\bullet$) activity recovered. (Reprinted from Ref. 19.)

and they have provided the incentive for continued work on this promising area for bioproduct recoveries.

## B.   Recovery of an Extracellular Enzyme from a Fermentation Broth

The true test for reversed micellar solutions as extractants in bioproduct recoveries is whether they are able to selectively extract proteins from real fermentation media.   This problem has been addressed by Rahaman and coworkers (15,16,19) using the recovery and purification of an extracellular alkaline protease (a detergent enzyme) from an untreated fermenation broth as the model system.   The results in Fig. 14 indicate that while over a range of pH conditions the total protein mass extracted varies little, possibly due to capacity limitations, the selectivity for the active component of the mixture is very strongly pH-dependent. This interplay between selectivity and capacity can be used to good advantage in the purification of the protease, as is illustrated in Fig. 15.   By increasing the aqueous/organic volume

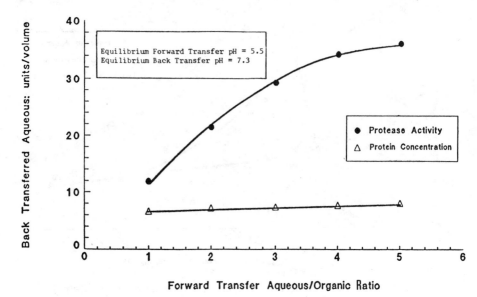

FIG. 15   Effect of aqueous feed/solvent ratio on extraction of an alkaline protease from a fermentation medium:   (Δ) total protein mass extracted, (●) total activity recovered.   (Reprinted from Ref. 19.)

ratio, the mass recovered per unit volume of extractant changes little, but the amount of active component extracted increases approximately fourfold over the range of conditions used in this work.   These results indicate that there is some competition for the micelles between the active and inactive proteins in the system.

Crosscurrent batch extraction studies at 1:1 aqueous/organic volume ratios yielded approximately 22% activity per stage, with about 10% of the protein mass recovered per stage, resulting in a 2.2-fold increase in protein-specific activity in the product phase.   While there is certainly room for optimization of the process to obtain greater yields and purification factors, these results are respectable, and are a clear demonstration of the potential for using reversed micelles with realistic protein mixtures of the type likely to be encountered in the biotechnology industries.

**TABLE 3** Isolation of Intracellular Enzymes from Intact Cells of *Azotobacter Vinelandii* Using Reversed Micelles[a]

| Enzymes | Conditions | Backward transfer medium | | | | | |
|---|---|---|---|---|---|---|---|
| | | KP$_i$ | | | Tris/HCl | | |
| | | Recovery (activity) % | s.a. U mg | Purification factor | Recovery (activity) % | s.a. U mg | Purification factor |
| Isocitrate dehydrogenase | CFE | 100 | 0.73 | 1 | 100 | 0.77 | 1 |
| | W$_o$ = 20 | 7.6 | 0.4 | 0.6 | 66 | 3.5 | 4.6 |
| β-Hydroxybutyric acid dehydrogenase | CFE | 100 | 0.034 | 1 | 100 | 0.056 | 1 |
| | W$_o$ = 20 | 26.8 | 0.066 | 1.9 | 92 | 0.35 | 6.2 |
| Glucose-6-phosphate dehydrogenase | CFE | 100 | 0.041 | 1 | 100 | 0.04 | 1 |
| | W$_o$ = 20 | 0 | 0 | — | 50 | 0.13 | 3.4 |

[a]Cells were disrupted by injecting 20 μl of a cell suspension (36 mg protein/ml) into 5 ml of a reversed micellar solution (W$_o$ = 6) consisting of a 0.2 M cetyltrimethylammonium bromide in hexanol/octane (1:9 vol/vol). The enzymes were subsequently recovered from the reversed micellar phase by backward transfer with a 0.8-ml buffered salt solution [0.5 M potassium phosphate (KPi) or Tris/HCl, pH 7.0]. Activities are normalized with respect to the activities found in cell-free extracts (CFE). s.a., specific activity.
*Source:* Ref. 22.

## C. Direct Recovery of Intracellular Enzymes

An interesting approach for the recovery of intracellular enzymes has been reported by Giovenco et al. (22), who injected a suspension of whole cells directly into a CTAB/isooctane reversed micellar solution. The detergent was instrumental in disintegratting the cell membranes, thereby releasing the cell enzymes, which were taken up by the reversed micelles. Subsequent recovery of the enzymes could be effected selectively, and with significant concentration factors, by appropriate selection of the aqueous stripping phase. They suggest that salt-detergent matching, with pH as an extra degree of freedom, is an important factor in the recovery of active enzyme, basing this conclusion on their observations with a number of different salt buffers. The buffer that appeared to be optimal was the bromide salt of the Tris buffer, which is structurally similar to CTAB. Their results are summarized in Table 3. It was claimed that $w_0$ could be used as a selection criterion in recovery of desired proteins and that the selectivity of the extraction should also be controllable by manipulating the surfactant/cosurfactant ratio.

A potential drawback to the use of this approach is the retention of the cell debris in the solvent phase, rendering the solvent unusable for further extractions. It has not yet been established, however, whether the solvent can be simply regenerated, possibly by inducing a phase change in the system by either temperature of chemical changes. If the solvent and surfactant recovery could be easily accomplished, this combined cell lysis/protein extraction process could be an important addition to the range of separation operations available for protein recovery.


## V.  PROCESS CONSIDERATIONS

The application of reversed micellar extraction of proteins to large-scale bioproduct recovery faces the same potential problems found in traditional extraction operations. It is important to attain high interfacial areas between the phases to ensure acceptable rates of mass transfer, and this is usually achieved by dispersing one phase in the other by mechanical means. Typical equipment types include simple mixer-settler systems, agitated countercurrent column-type contactors, and centrifugal units of the Podbielniak and Westfalia designs (23). The column and centrifugal continuous contacting devices typically have low-stage efficiencies owing to channeling problems and incomplete mixing

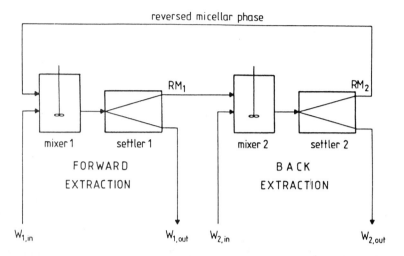

FIG. 16  Mixer settler configuration for the continuous extraction of α-amylase using a TOMAC/isooctane solution. (Reprinted from Refs. 5 and 7, with permission.)

of the two phases.  They also have a rather restricted range of operating conditions, in particular relative flow rates, as it is essential to avoid entrainment of one phase by the other, a condition known as "flooding."  Moreover, the presence of high levels of surfactants in the system could lead to the formation of relatively stable emulsions, thus exacerbating the flooding problem.  Phase disengagement can be enhanced, however, by one of the centrifugal contactors.  Indeed, such contactors have been used successfully for decades in the recovery of antibiotics from both clarified and whole-broth fermentation media, and problems associated with emulsion formation have, by and large, been overcome.

Of these contacting types, only the use of a mixer-settler system has been demonstrated.  Dekker and coworkers (5,7) were able to extract α-amylase from an aqueous solution into a TOMAC/ isooctane extractant and then to strip out the enzyme into a second, buffered aqueous product stream using the extraction configuration shown in Fig. 16.  The results are shown in Fig. 17, where the concentrations of active enzyme in the phases leaving the two mixer-settlers are given as a function of time from the start of the experiment, expressed in terms of the recirculation

relative α-amylase activity

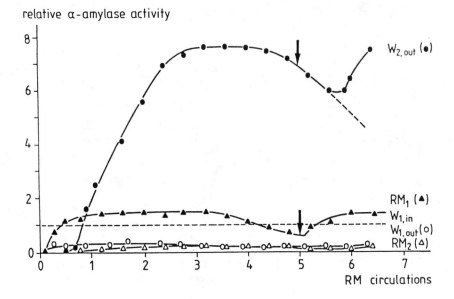

FIG. 17   Performance of the continuous extraction system.  Concentration of active α-amylase in the indicated phases versus the
number of recirculations of the reversed micellar phase.   Addition
of makeup surfactant indicated by arrow.  (Reprinted from Ref.
5, with permission.)

rate of the reversed micellar solution between the extractor and
the stripper.   Steady-state operation was attained after about two
recirculation times, but the extraction efficiency began to decrease
after approximately 3.5 recirculations, attributed to a gradual
loss in surfactant to the aqueous phase ($<15\%$ per pass).   On
replenishment of the lost surfactant, the extraction efficiency
was returned to its original value.   Under the operating conditions selected, about 45% of the enzymatic activity was recovered
in the product stream, with an eight-fold concentration factor.
Inactivation of the enzyme accounted for 30% of the activity, the
remaining 25% being retained in the feed phase.   These results
are encouraging and point to the feasibility of using reversed
micelles in the large-scale recovery and concentration of bioproducts.
There is, of course, room for significant optimization of this
process and it is remarkable that this level of success has been
attained so early in the development of this process.

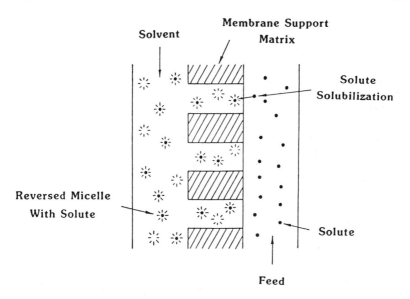

FIG. 18    The concept of a membrane extractor.


A particular disadvantage of agitated systems is the intense
mixing required to effect intimate contact between the two phases
and the attendant problems of phase disengagement and possible
emulsification.    These and other disadvantages can be avoided to
a large extent using microporous membrane contactors.    In these
systems, depicted schematically in Fig. 18, the phases are
passed on either side of the membrane.    The membrane is preferentially wet by one of these phases, which wicks through and
fills the membrane pores.    Consequently, the interface between
the wetting and nonwetting liquids tends to move to the nonwetting surface of the membrane where it can shear off to form
an emulsion.    This emulsion formation can be avoided by applying
a higher static pressure to the nonwetting fluid to push the
liquid-liquid interface back into the membrane pores where it will
not be subjected to shear.    Surface areas per unit volume in
hollow fiber modules are usually much higher than those attainable
in agitated systems, and this more than offsets the reduced overall mass transfer coefficient for these membrane contactors.
Thus, it is evident that hollow fibers could offer significant advantages over the usual agitated contacting devices.    Moreover,

in hollow fibers, the two fluid flows are almost completely independent and therefore flooding, loading, and channeling are
essentially avoided.  Consequently, countercurrency in operation
can be observed using highly disparate flow rates of the two
phases, while still retaining extremely high interfacial areas,
even if these phases have similar (or identical!) densities.  Such
operating conditions cannot be achieved in traditional agitated
systems.

It has been shown by three groups that membrane extractors
may be used in the reversed micellar extraction of proteins.
The most complete study is that of Dahuron and Cussler (24)
using the AOT/isooctane system to extract cytochrome c and
α-chymotrypsin from buffered aqueous solutions.  They were
able to chracterize their results in terms of traditional mass
transfer correlations for flows inside and outside the fibers,
respectively, and for the diffusional resistance within the pores
of the membranes.

Dekker et al. (8) and Czupryna et al. (25) also demonstrated
the feasibility of using membrane extractors for reversed micellar
extraction of proteins, the latter authors showing, in addition,
the separation of protein mixtures using this approach.

In none of the membrane extraction studies mentioned above
was the protein recovered from the organic phase.  An alternative configuration can be employed in which the extraction and
stripping are accomplished simultaneously in a liquid membrane
apparatus.  The pores of the membrane are impregnated with
the organic solvent/surfactant solution and the feed and stripping
solutions are flowed on either side of the membrane.  The protein
is extracted on the feed side of the membrane by the reversed
micelles and transferred by diffusion to the stripping solution
side of the membrane.  Here the protein is surrendered to the
stripping solution and the surfactant micelles, now empty, and
possibly rearranged, return to the feed side of the membrane to
pick up more protein.  The advantages of this approach are that
the protein product is both extracted and stripped in the same
piece of equipment, and that solvent and surfactant inventories
are drastically reduced because the reversed micelles simply act
as shuttles for the protein, and not as repositories for the biomolecule.  Thus, the capacity limitation of the solvent is not a
significant factor.  The utility of liquid membranes has been
demonstrated by Czupryna et al. (25).

## VI. CONCLUSIONS

Reversed micellar solutions have demonstrated potential as effective extractants for proteins and other bioproducts from aqueous fermentation media. ,There is a great deal of latitude in manipulating solution conditions to optimize selectivity, which can be strongly influenced by pH, salt and surfactant concentration and type, solvent type, and temperature. Moreover, incorporation of small amounts of a biospecific affinity ligand can enhance dramatically the effectiveness of the extraction and recovery of the trageted protein.

While some attention has been given to the processing requirements for operating reversed micellar extraction on a continuous, large-scale basis, there is still much to be done on the optimization of these processes. An understanding of the basic science behind the solubilization process will of crucial importance in selecting optimum operating conditions.

## REFERENCES

1. P. L. Luisi, F. J. Bonner, A. Pelligrini, P. Wiget, and R. Wolf, *Helv. Chim. Acta, 62:* 740 (1979).
2. P. L. Luisi, V. E. Mire, H. Jaeckle, and A. Pande, in *Topics in Pharmaceutical Sciences* (D. D. Breimer and P. Speiser, eds.), Elsevier, Amsterdam, 1983, p. 243.
3. P. Meier, E. Imre, M. Fleschar, and P. L. Luisi, in *Surfactants in Solution*, Vol. 2 (K. L. Mittal and B. Lindman, eds.), Plenum Press, New York, 1984, p. 999.
4. K. Van't Riet and M. Dekker, *Proc. 3rd Eur. Congr. Biotechnol., 3:* 541 (1984).
5. M. Dekker, K. Van't Riet, S. R. Weijers, J. W. A. Baltussen, C. Laane, and B. H. Bijsterbosch, *Chem. Eng. J., 33:* B27 (1986).
6. M. Dekker, J. W. A. Baltussen, K. Van't Riet, B. H. Bijsterbosch, C. Laane, and R. Hilhorst, in *Biocatalysis in Organic Media* (C. Laane, J. Tramper, and M. D. Lilly, eds.), Elsevier, Amsterdam, 1987, p. 285.
7. M. Dekker, K. Van't Riet, J. W. A. Baltussen, B. H. Bijsterbosch, R. Hilhorst, and C. Laane, *Proc. 4th Eur. Congr. Biotechnol., 2:* 507 (1987).

8.  M. Dekker, K. Van't Riet, J. M. G. M. Wijnans, J. W. A. Baltussen, B. H. Bijsterbosch, and C. Laane, *Proc. Int. Congr. on Membranes*, Tokyo, 793 (1987).

9.  K. E. Goklen and T. A. Hatton, *Biotechnol. Prog.*, *1*: 69 (1985).

10. K. E. Goklen and T. A. Hatton, *Proc. Int. Solvent Extraction Conf. '86*, *3*: 587 (1986).

11. K. E. Goklen, Ph.D. dissertation, Massachusetts Institute of Technology, Cambridge, Mass., 1986.

12. K. E. Goklen and T. A. Hatton, *Sep. Sci. Technol.*, *22*: 831 (1987).

13. T. A. Hatton, in *Ordered Media in Chemical Separations* (W. L. Hinze and D. W. Armstrong, eds.), ACS Symp. Ser., Vol. 342, 1987, p. 170.

14. J. M. Woll, M.S. thesis, Massachusetts Institute of Technology, Cambridge, Mass., 1987.

15. J. M. Woll, A. S. Dillon, R. S. Rahaman, and T. A. Hatton, in *Protein Purification: Micro to Macro* (R. Burgess, ed.), Alan R. Liss, New York, 1987.

16. R. S. Rahaman and T. A. Hatton, *Proc. 2nd Int. Conf. on Separation Technol.*, Engineering Foundation, Washington, D.C., 1987 (in press).

17. J. M. Woll and T. A. Hatton, *Bioproc. Engng.* (in press).

18. J. M. Woll, T. A. Hatton, and M. L. Yarmush, *Biotechnol. Prog.* (in press).

19. R. S. Rahaman, J. Chee, J. Cabral, and T. A. Hatton, *Biotechnol. Prog.* (in press).

20. M. E. Leser, G. Wei, P. L. Luisi, and M. Maestro, *Biochem. Biophys. Res. Commun.*, *135*: 629 (1986).

21. S. D. Flanagan and J. H. Barondes, *J. Biol. Chem.*, *259*: 1484 (1975).

22. S. Giovenco, F. Verheggen, and C. Laane, *Enzyme Microb. Technol.*, *9*: 470 (1987).

23. T. C. Lo, M. H. I. Baird, and C. Hanson (eds.), *Handbook of Solvent Extraction*, Wiley, New York, 1983.

24. L. Dahuron and E. L. Cussler, *AICHE J.*

25. G. Czypryna, R. Levy, H. Gold, and T. A. Hatton (unpublished results).

# 4
# Novel Separations Using Aphrons

FELIX SEBBA  Department of Chemical Engineering, Virginia
Polytechnic Institute and State University, Blacksburg, Virginia

SYNOPSIS

Aphrons are globules of phases encapsulated in a soapy film. A soap bubble is a giant gas aphron; colloidal gas aphrons comprise gas bubbles that are of micrometer size, and colloidal liquid aphrons have the gas replaced by a liquid that is immiscible in water and can be submicrometer in size. Presenting an enormous interfacial surface area, these systems have potential in separation processes. The properties of these systems are discussed in this chapter. Ion flotation, which uses small gas bubbles and surfactants for buoying ions to the surface, is improved by using colloidal gas aphrons as in precipitate flotation. These can also be used to float finely divided solids that are too small to be collected by conventional flotation as well as fine oil droplets. The bubbles are so small that they can be entrained in flocs conferring buoyancy (bubble-entrained floc flotation). Predispersed solvent extraction uses oil core aphrons, eliminating the need for a mixer-settler unit.

I. INTRODUCTION

As a group of substances, surfactants play an important role in separation procedures for two reasons. First, they can serve as energy barriers, thus enabling fluid media to be stabilized in the form of very small globules thereby exposing an enormously increased interfacial area, where transfer from one phase to another can occur very rapidly. Second, because of their amphipathic nature, they tend to adsorb at interfaces. Thus, they can themselves act as collectors in flotation procedures. As will be shown, some techniques combine both these attributes in one procedure, thus producing very powerful separation techniques.

Because this chapter has as its theme some of the uses of surfactants in separation procedures, it is important to recognize that there are two distinct groups of surfactants. One group consists of surfactants that are significantly soluble in water. When dissolved they will produce:

1. Anions, such as laurates or alkyl sulfonates
2. Cations, such as alkyl ammonium or alkyl aminium
3. Uncharged water-soluble alkyl polyethylene alcohols of high HLB

The second group consists of polar but uncharged molecules which have very low solubility in water, if any, but are soluble in nonpolar solvents such as hydrocarbons. Using the terminology customary in emulsion technology, such solvents will be referred to here as oils. These surfactants fall into three categories:

1. Fatty acids such as stearic or lauric acid.
2. Bases such as the fatty amines, e.g., dodecylamine
3. Neutral molecules such as alkyl polyethylene alcohols of low HLB, or fatty alcohols such as cetyl alcohol

There is another class of surfactant comprising those that have been especially tailored for separation processes. One such group consists of molecules that serve as chelating agents, but with a fatty alkyl radical incorporated to produce the required surfactant properties. Another such group consists of mercaptans, i.e., thio-alcohols to which a similar fatty radical has been attached.

## II.  APHRONS

The recent developments in colloidal systems based on phases encapsulated in a thin soap film, and known as aphrons, opens up many new opportunities for separation techniques. Thin soap films have considerable mechanical strength due to the thermo-dynamically induced Gibbs elasticity produced by the water-soluble surfactant that is needed for the production of a soap film. For that reason phases encapsulated in a soap film are surprisingly stable. Two types of aphrons concern us here: gas aphrons and liquid core aphrons.

The simplest example of a gas aphron is a free-floating bubble, which is essentially a giant gas aphron. However, it is when the bubbles are very small that useful properties appear. When the bubbles are from 25 to 100 μm in diameter and dispersed in water they are known as colloidal gas aphrons or CGA. They are similar to kugelschaums, which are foams comprising extremely small, spherical bubbles (1). The small size of the bubbles, which gives them colloidal properties, and their being perfect spheres results in a fluid that flows as easily as water. The facility with which these bubbles can be pumped from one vessel to another produces a system with considerable potential in a remarkable diversity of applications, one of which is separations.

An oil core aphron is a globule, usually of micrometer or submicrometer size, of an oil phase, also encapsulated in a soapy film. Oil, in this connection, means a liquid that is immiscible in water. A concentrated aggregation of such aphrons is known as a polyaphron. Just as a conventional foam is an aggregate of gas aphrons, a polyaphron is similar except that, instead of a gas in each cell, there is an oil phase. Systems of that type are known as biliquid foams. The preparation and properties of polyaphrons have been described by Sebba (2).

The production of a polyaphron needs a surfactant in the aqueous phase to stabilize the encapsulating soap film, and often a small quantity of an oil-soluble surfactant in the oil phase. This is because the spreading of the oil on the water as a thin film is an essential part of the method for making the polyaphron. The concentration of this surfactant can be used to control the size of the aphrons:  the higher the concentration, the smaller the aphron. Usefully, these can range from 1 to 100 μm. Poly-aphrons are best characterized by the phase volume ratio (PVR), which is the ratio by volume of discrete phase to continuous phase. Polyaphrons have proved to be very stable. Some composed of 100 parts kerosene and 5 parts water, PVR 20, water being the continuous phase, if properly prepared, can last for years without deterioration. Thus, if used for solvent extraction, the solvent can be predispersed as a polyaphron and stored until it is needed.

## III.  COLLOIDAL GAS APHRONS

### A.  Preparation of CGA

Though there are a number of ways to prepare a CGA, by far the most satisfactory is the spinning disk CGA generator (3). This requires a horizontal disk, which spins very rapidly (above 4000 rpm) and is positioned about 2 cm below the surface of the surfactant solution, as shown in Fig. 1. The disk is mounted between two vertical baffles, which could be made of rigid plastic, and which extend well above the surface of the solution. To avoid wobble, the shaft which carries the disk should be support-ed by bearings mounted below the stirring motor. The motor should be mounted on a sturdy support. If these precautions are taken, the disk can be spun at 6000 rpm continuously for very many hours without any untoward effects. If the disk rotates at less than 4000 rpm, no CGA is formed, but once a critical speed is reached, waves are produced on the surface.

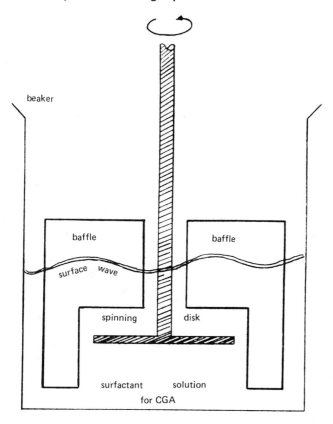

FIG. 1   Spinning disk CGA generator.

These beat up against the baffles and, having nowhere else to
go, reenter the liquid at the baffles.   It is believed that the
reentering liquid carries with it a thin film of gas that becomes
sandwiched between the liquid and the baffle.   Such a thin film
of gas is unstable and will break up into minute droplets of gas
encapsulated by the soapy shell, i.e., minute gas aphrons.   The
speed at which these can be made is remarkable.   Three liters
of CGA can be made in a few seconds.   The unit has been run
for many years without malfunction.   If production of CGA is
needed in larger quantities, there seems no reason why several
such generators should not be used in parallel.   The principal

outlay is for the stirring motor. It has been estimated that the
cost of electricity for 10,000 liters of CGA would be less than
the cost of 1 KW-hr of electricity.

## B.  Properties of Small Bubbles

There are a number of remarkable properties of very small
bubbles that could only be observed as a result of the development of CGA.  This is because they rise only very slowly under
the influence of gravity, so that it became possible to devise a
system for observing them under high-power microscopy.  Perhaps the most unexpected observation is that micrometer-sized
bubbles never coalesce, provided they are completely surrounded
by water.  When they collide the momentum is not enough to
break the encapsulating soap film.  They therefore simply bounce
off one another or roll on one another like ball bearings.  They
are, of course, under some pressure because of the excess
pressure inside a bubble as given by the Laplace equation, which
shows that the excess pressure inside a free-floating gas bubble
in air is $4\gamma/r$, where $\gamma$ is the surface tension of the water and r
is the radius of the bubble.

It is seen, therefore, that the excess pressure is inversely
proportional to the radius, so that the smaller the bubble, the
higher the pressure.  Thus, the minute gas aphrons show little
distortion when they touch, unlike the situation in polyederschaums
(conventional foams with polyhedral cells), when parallel lamellae
are common.

A soap bubble that is 1 μm in radius, where the surface
tension is 25 dynes/cm, will have excess pressure inside it of
approximately 1 atm.  This pressure is compensated for by the
tension in the encapsulating soap film.  It follows that the
smaller bubbles will tend to diminsh in size by transferring their
gas to the larger ones, which would have a lower excess pressure
and would accordingly grow larger.  What is interesting is that
the bubbles do not necessarily have to be in contact, so there
has to be a diffusion gradient between small bubbles and the
larger ones, as well as between the bubbles and the upper surface of the water, where there would be no excess pressure, as
illustrated in Fig. 2.

## C.  Minimum Size of Bubbles

This excess pressure imposes a lower limit to the size of the
bubbles.  In practice it has not been found possible to make
CGAs that have bubbles smaller than 25 μm in diameter.  Any

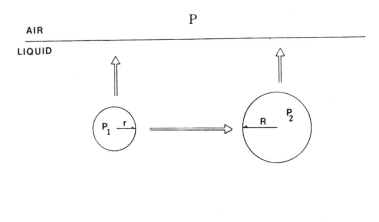

$$P_1 = P + \frac{4\gamma}{r} \qquad\qquad P_2 = P + \frac{4\gamma}{R}$$

If $R > r$,     $P_1 > P_2 > P$

FIG. 2   Diffusion from bubbles.

that are smaller disappear very rapidly. Many attempts have
been made to stabilize smaller bubbles, such as by increasing
the viscosity of the water, but any reduction of the rate of
diffusion does not have any significant effect on the final concentration of small bubbles. If a CGA is put through a mechanical
homogenizer, bubbles smaller than 25 μm are produced, but these
too are shortlived. Ultrasonic disintegration too has only a
transitory effect. Reluctantly, it has to be accepted that CGAs
with bubbles smaller than 25-μm diameter are not possible (4).

When newly formed gas aphrons are examined under the
microscope they are seen to be strikingly uniform in size, but
it is not long before a distribution in size becomes apparent and
numbers of small bubles of less than average size are observed.
However, these are seen to diminish very rapidly and eventually
vanish. Thus, a CGA that is more than a few minutes old will
be seen to be composed largely of aphrons of about 25 μm in
diameter and above. There is some uncertainty as to the actual
dimensions of the bubbles, since because of the complex optical

effects it has not been possible to distinguish the actual boundary of the shell of the bubble from an optical illusion.

## D. Creaming of CGAs

CGAs show the same tendency as emulsions to cream, i.e., have
the lighter oil droplets rise slowly to the surface due to their
being less dense than the water in which they are dispersed.
However, because of the much greater buoyancy of the gas
aphrons due in part to their larger size (hence less resistance
to movement), but also because of the much lower density of
the encapsulated gas, creaming of a CGA is much faster than
the creaming of emulsions. On standing, a CGA will, in 10–15
min, separate into a conventional foam floating on clear water.
However, the creaming can be delayed by stirring the CGA at
such a rate that the lateral movement conveyed to the bubbles
is greater than the upward buoyancy due to gravity. In this
way a CGA can be maintained for an hour or more. The same
effect is produced when the CGA is made to travel along a tube.
It is for that reason that the CGA can be pumped to the place
where it is to be used, one of the very useful properties of a
CGA. However, it must be borne in mind that the low viscosity,
which enables CGA to be pumped so easily, only applies when
the bubbles are perfect spheres. When creaming occurs, the
bubbles become more crowded so that they become distorted and
the viscosity increases. It is therefore important that the CGA
be drawn from a region below that of any creamed bubbles.
For ease of pumping, a CGA should have a PVR below 2.

This creaming provides for one of the important applications
of CGA in separations, namely, the buoying of small particles to
the surface, i.e., flotation. Once the bubbles reach the surface,
they become more crowded and change from a kugelschaum to a
polyederschaum, with thin lamellae through which water drains
until the lamellae get so thin that they burst. This means that
the cells become larger, so that the foam eventually collapses.

## E. Properties of CGA

Perhaps one of the most useful properties of a CGA depends on
the structure at the interface between the bubble and the water
in which it is dispersed. Figure 3 shows how a gas aphron is
constituted at the molecular level. It should be noted that the
encapsulating soap film has an inner as well as an outer surface.
Though these surfaces have surfactant monolayers adsorbed on
them, it must be remembered that these are expanded monolayers

Outer surface of shell

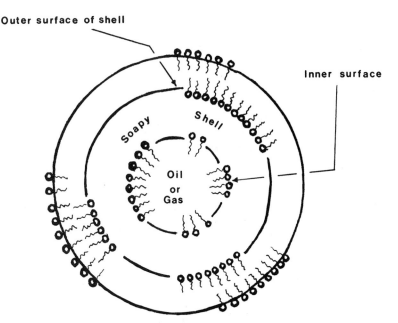

Bulk water containing surfactant

FIG. 3    Structure of aphrons.

so that the relative area of the surface occupied by monolayers
is small, the rest being water. However, this water has different properties from bulk water, probably being more hydrogen-
bonded (5). For that reason, it is quite permissible to consider
this water as being a different phase from bulk water. This
phase has oriented surfactant molecules at the surface that are
hydrophilic pointing inward and hydrophobic outward. This is
easy to visualize for the surface in contact with the gas, but not
so easy to understand for the surface exposed to bulk water.
However, if it were a free-floating gas bubble, there would be
no problem in recognizing that the surfactant will be hydrophobic
facing outward. As the structure of the bubble is still maintained
when it is dispersed in water, it too must have its surfactant
hydrophobe facing outward.

What about the bulk water with which it is in contact? If
the soap shell is behaving as a separate phase from bulk water,
there must be an interface between that phase and the water.
The Gibbs adsorption isotherm must then apply, and there must
be a higher concentration of surfactant at that surface of the
bulk phase than in the interior. That surfactant must also be
orientated and, as the surface of the shell is hydrophobic, its
orientation too must be hydrophobic facing the soapy shell.
This provides a region where a hydrophobic globule will be
comfortable, and it has been shown that small globules of oil as
well as small particles tend to adhere to a gas aphron. These
captured globules can thus be buoyed to the surface by the
rising bubbles. In some respects the interface is acting as a
two-dimensional micelle that can solubilize nonpolar compounds
because of their hydrocarbon-like interior, except that here
it is not solubilization but physical adherence that is operating.


## IV.  ION FLOTATION

The principle of ion flotation depends on the fact that if a gas
cavity is created by sparging in an aqueous solution that contains
a dissolved surfactant, the surfactant will adsorb at the gas-
water interface with the headgroup orientated toward the water.
If it carries a charge it may form an ionic compound with an
oppositely charged species in the solution. As the cavity will
quickly rise to the surface of the water due to its lower density,
it will buoy the attracted ion to the surface. As the bubble will
soon burst at the surface, this will tend to concentrate that ion
at the surface, particularly if its compound with the soap is
virtually insoluble. It should be noted that, in this type of
separation, it is adsorption at a hole that is important and not
at a bubble with two surfaces as in CGA flotation. This is a
separation process that is dependent on the charge at the polar
head of the surfactant, although there are a few instances where
an uncharged amine head can be used because the lone pair of
electrons on the nitrogen atom can form coordination complexes.
An advantage of ion flotation as a separation technique is that
it can float ions sequentially, i.e., it can be made selective.
Ion flotation was first described in a short note by Sebba (6).
A fuller account is given in the monograph *Ion Flotation* (7). A
more recent account occurs in an article by Grieves (8).

## V.  PRECIPITATE FLOTATION

Although ion flotation has the capability of concentrating ions
from extremely dilute solutions, even parts per billion, it suffers
from the disadvantage that there is a stoichiometric relationship
between colligend ion (that which is being collected) and the
surfactant ion.  This makes the technique expensive and it has
not yet found favor on the industrial scale.  That criticism,
however, does not apply to the allied technique of precipitate
flotation.  In precipitate flotation, a precipitate, usually colloidal,
is formed at the same time as the surfactant together with sparged
gas are added.  This applies particularly well when the precipitate
is an hydroxide such as aluminum hydroxide or ferric hydroxide.
These precipitates carry a charge that is usually the same as
that of the precipitating ion that is in excess.  If the surfactant
carries a charge that is opposite to the charge of the ion, the
surfactant will attract the precipitate as well as adhere to the
sparged cavity, so the aggregate will be buoyed to the surface.
However, in this case, there is no longer a stoichiometric relationship between colligend and surfactant.  In fact, the ratio can
now be as high as 1000 of colligend to 1 of surfactant, which
makes the process much more economical.
    It is essential for successful flotation, using either ion or
precipitate flotation, that the sparged bubbles be so small that
they do not cause much disturbance of the froth layer on the
surface.  CGA bubbles are small enough to be used in ion flotation, but they have two other advantages in that they can be
produced away from the flotation cell and can be accurately
metered into the flotation cell.  It is anticipated that development
of CGA will give renewed vitality to these two older techniques.
Ciriello et al. (9) have successfully removed chromium, nickel,
and zinc from electroplating effluents by ion flotation using CGA.
Shea and Barnett (10) removed dyes from effluents by ion flotation.  In many instances, when CGA is used there is a residual
turbidity in the water due to the fact that the smallest bubbles are
slow to rise.  The water can be quickly clarified by a precipitate
flotation of aluminium hydroxide.  A small quantity of aluminum
sulfate is added to the solution, the pH is adjusted to about 5,
and a CGA made of sodium dodecyl benzenesulfonate is introduced.
A mixture of aluminum soap and aluminum hydroxide is floated to
the surface and this entrains all the suspended matter, leaving
a perfectly clear water.

## VI.  FLOTATION OF FINELY DIVIDED SOLIDS FROM WATER

This section will consider only the flotation of dispersed solids
by their adherence directly to the bubbles.  A more general
method, which combines flotation with flocculation, will be discussed in Sec. VIII.  The outer surface of the shells, which en-
capsulate the CGA bubble aphrons, has a surface elasticity as
well as some degree of hydrophobicity.  Fine solids that meet
this surface appear to have a tendency to stick in it and can
thus be buoyed to the surface of the water in which they had
been dispersed.  Unlike ion and precipitate flotation, which
depend on the attraction of opposite charges, this type of flotation apparently does not have any charge relationship.  Therefore, uncharged particles can be floated by a CGA, which had
been made using surfactants of any charge, or even nonionic
surfactants.  Barnett and Lin (11) showed that the debris floating
in fish culture ponds can be effectively removed by floating
with CGA.  Barnett (12) also successfully floated finely dispersed
cellulose in the same way.  It is somewhat surprising that this
material, which is so hydrophilic, should leave an aqueous environment and attach to the soap bubble.  Possibly the highly
hydrogen-bonded nature of the aphron shell has something to do
with this.

## VII.  FLOTATION OF DISPERSED OIL DROPLETS FROM WATER

Flotation of dispersed oil droplets from water has the potential
for developing into one of the most important applications of CGA
as removal of fine oil dispersions from wastewaters often presents
a difficult problem.  The ability of CGA bubbles to collect oil
dispersions depends on the slight hydrophobicity of the encapsulating soap film surrounding the bubbles.  The oil droplets tend
to adhere to the outside of this film locating between it and the
bulk water.  Because generally the oil droplets are considerably
smaller than the gas bubbles, a large number of droplets can be
accommodated on each aphron.

If the radius of the aphron is R μm, and the radius of the
oil droplet is r μm, assuming the surface of the aphron is completely covered by spheres of oil, the number of drops of oil
accommodated on each bubble would be $\pi(R/r)^2$.

If the volume of air per liter of CGA is V ml, the volume of oil that could be buoyed to the surface would be $Vr/R$. If 1 liter of CGA contains 60% of air, i.e., V = 600 ml, and if R = 12.5 $\mu$m and r = 1 $\mu$m, then the volume of oil that could be removed by 1 liter of CGA would be 48 ml.

As the concentration of oil in a wastewater containing an oil dispersion would be quite small, a considerable volume of water could be cleaned of oil using a comparatively small volume of CGA.

This calculation is somewhat oversimplified as the adhering droplets are not in fact spheres but are more like cusps. This is because most oils will contain some surfactant impurity that will tend to make the drops spread on the shell interface, i.e., become lenses. This will reduce the number that can be accommodated somewhat, but the scavenging effect will still be considerable.

Often, but not always, when the CGA bubbles break through the water-air interface, the bubbles become destabilized by the presence of the oil so that they break, the oil coalesces, and there is very little buildup of foam. If there is any foam, it is probably due to the presence of fine solid particles. Even so, the foam eventually breaks, the oil droplets coalesce, and a layer of oil floating on the surface is produced.

There are as yet no quantitative data on the efficiency of the removal but it has been tried qualitatively on several different samples of wastewater contaminated with oil. After a CGA flotation, what started as very turbid water became perfectly clear. The method was also tried by this author on the effluent from a tar sands extraction plant. This had a suspension of very fine oil droplets that were so difficult to remove that the effluent had to be pumped to, and retained in, a waste slimes dam. The CGA in a very few minutes clarified the water so that it could undoubtedly have been returned to the circuit.

Oils other than hydrocarbons can be removed in a similar way. Nitrobenzene has been floated, and it has been privately communicated that limonene has been recovered from the wastewaters from a citrus-processing plant. As limonene has a commercial value, this is a case where clean-up of a wastewater could be a profitable process, as the cost of producing the CGA is so low. It should be noted that no collector is needed for such flotation, the spreading pressure of the oil being enough to promote adherence to the bubbles. The main cost is the cost of the surfactant needed to make the CGA, but as the concentration needed is only about 0.3 g/liter and the cheapest surfactant can be used, this cost is negligible.

## VIII.  BUBBLE-ENTRAINED FLOC FLOTATION

Bubble-entrained floc flotation (BEFF) is the name given to the process in which, instead of the solid adhering to the outside of the bubble, the solid is produced by flocculation in such a way that the CGA bubbles are entrained in the floc, thus making it buoyant.  This becomes possible partly because the CGA bubbles are so small and partly because the bubbles can be introduced before flocculation and, being so small, still remain present while flocculation, which is a somewhat slow process, proceeds.  This was first described in Auten and Sebba (13) and further developed in Alexander et al. (14).  There is photographic evidence to show that bubble entrainment does in fact occur.  Photographs taken in a cell specially designed for the purpose, which enabled the flocs to be seen under a microscope, clearly show the bubbles entrapped in the floc.  The method shows promise on the industrial scale, especially for water treatment plants, for the separation of flocs because it eliminates the need for expensive filtration procedures, which are sometimes very slow where gelatinous flocs are concerned.

The method has been successfully demonstrated on a bench scale for the removal of the colloidal clays from the so-called Florida phosphate slimes.  The phosphate ore in Florida originated as a result of marine deposition of biological debris.  Associated with the phosphate is a considerable quantity of finely divided clays.  In the processing of the ore, the first stage is the removal of much of this clay by simple washing with pressured jets of water.  This is in order to avoid waste of expensive reagents by adsorption on the clay materials during the froth flotation process, which is used to upgrade the phosphate ore. The washwater contains the clay in the form of a fine suspension that will not settle completely even after standing for several years.  It thus presents a considerable environmental embarrassment, and it has to be stored in large slime dams that occupy a considerable acreage of valuable land in an unproductive way. There is also a considerable quantity of water wasted.

It was found that it is possible to float the clays very quickly and efficiently using the bubble-entrained floc flotation technique. One procedure for doing this is to add suitable flocculants such as aluminum sulfate, with pH adjustment to 5.5, or to add cationic

starches or other polyelectrolytes to the slimes, which would
sometimes need further dilution before flocculating. The flocculant should be cationic as the slimes colloids are negatively
charged. At the same time a CGA made with an anionic surfactant is introduced into the suspension carefully to avoid dis-
turbing the froth that builds up on the surface of the water.
The clays flocculate entraining the small bubbles in sufficient
amount to make them buoyant in the water. They are thus taken
to the surface where they are retained in the froth as long as
it persists. Surprisingly, the solid content of the froth may be
less than that in the slimes itself. There is an important difference, however. In the slimes much of the water is held to the
clays by chemical forces, which is the reason why the slimes
are so difficult to dewater. In the froth, most of the water is
held by capillary forces in the interstices between the clay
particles. Such water is more easily removed and, because of
the large surface area introduced by the bubbles, such a froth
will dry in the air in a day or so, producing a fine powder of
dried clay. The froth has another property. Whereas the
slimes are very difficult to filter, the froth produced by bubble-
entrained floc flotation filters very easily and can be successfully
dewatered in that way. This is because the reaction of the
cationic flocculant and the anionic surfactant produces an hydrophobic compound.

The flocculating agent can be introduced in the CGA itself,
the bubbles apparently nucleating the floc formation. If that is
done it becomes possible to achieve a separation, as, for example,
between the clays and residual apatite in the Florida slimes. In
such cases care has to be taken that there is no repulsion between the charges on the flocculating agent and the charges on
the bubbles that may be produced by the surfactant. The use
of nonionic surfactants proved to be useful in such cases. As
the clay particles carried a negative charge, it was necessary
to stabilize the CGA by using a cationic surfactant. This had
the additional advantage that it served as a flocculant for the
clays. However, the surfactant chosen had to be one that did
not flocculate the apatite meant to be separated. Screening
tests using a 500-ml batch cell established that dodecylaminium
chloride was a suitable flocculant. It tends to hydrolyze, however, so it is not an ideal surfactant for generating CGA. For
that reason it was found preferable to use a mixture of surfactants for generating the CGA, the amine to serve as a flocculant
and a nonionic surfactant to provide a better CGA. For this
purpose a suitable surfactant was the polyethylene glycol ether

of secondary alcohol, Tergitol 15-S-9, made by Union Carbide
Corporation. A satisfactory formulation for the CGA solution
was 0.1 g/liter of dodecylaminium chloride and 0.25 g/liter of
Tergitol 15-S-9.

Under these conditions a considerable amount (up to 30% of
total solids) of finely divided phosphate rock still present in the
washwater slimes was not floated and fell as a sediment to the
bottom of the flotation cell. This demonstrates that given the
correct conditions, bubble-entrained floc flotation can be made
selective. In this separation the solid particles were less than
1 μm in size. Details are given in Alexander et al. (14), but
as a number of interesting considerations have emerged from
this work, some of it will be described here.

As the process had to be one that would be suitable for a
large-scale operation, a continuous system was developed in
which the flotation cell was a horizontal, V-shaped trough, in
which the feed and CGA entered at one end, and the froth
containing the clays was run off at the far end over a launder.
The clear water was removed at the same end as an underflow.
The unfloated apatite dropped out as a sediment close to the
point of entry. A laboratory trough for such flotation is shown
in Fig. 4. In order to facilitate collection of the sediment, the
V was truncated to a flat base as the sediment otherwise was
difficult to collect. As one of the objectives was to produce
water that could be reused, a baffle was introduced about one-
third of the way along the trough. This was to reduce mixing
of the raffinate with the froth because of the turbulence introduced by the momentum of the entering feed and CGA.

The consistency of the froth, which carries the floated clays,
is very dependent on the nature of the surfactant and the
quantity of solid entrained. The solid clays when incorporated
into the froth tended to make it almost pasty, so that it did not
move of its own accord. On the other hand, if conditions were
so arranged that the froth was mobile enough to flow easily
toward the launder, there was a tendency for the froth to release some of the entrained solids so that they fell back into
the water in the trough contaminating both the water and the
sediment. It was therefore decided that it was better practice
to have a more adherent froth, albeit it did not flow so well,
and impel it toward the launder with a series of motor-driven
paddles that were adjusted so as to be just above the interface
between the froth and the water. The paddles were moved at
a speed of approximately 120 cm/min.

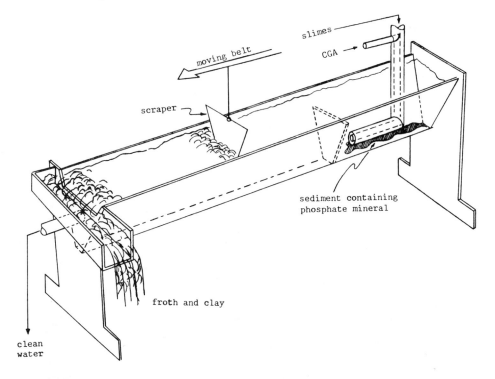

FIG. 4   Trough for bubble-entrained floc flotation.

When the separation of two types of solids, which are very
finely subdivided, is required, one of the problems is that of
adherence of the two different solid particles.  Clearly, the more
dilute the system, the less chance there is of that happening.
These considerations apply in bubble-entrained floc flotation.
The more dilute the feed, the less chance there is of the unflocculated sediments being caught up in the floc.  The effect
of dilution was studied by comparison of the flotation of slimes
diluted with tap water in ratios of 1 of slimes to 1, 3, and 7 of
water as well as undiluted slimes.  In the case of the undiluted
slimes, the fallout of all the solids from the froth was so great
that the performance was poor.  The clearest raffinate was obtained with a 1:7 dilution, but a 1:3 dilution gave a good phosphate mineral recovery and it would seem that this dilution would

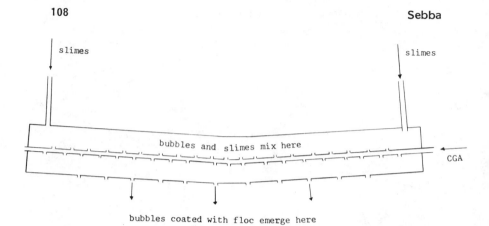

FIG. 5   Mixing system for CGA and slimes.


provide a good compromise between optimum clarity of the raffinate and maximum phosphate recovery.

   Of various methods tested for ensuring the best method for
mixing CGA and slimes, by far the most successful method was
the use of two concentric tubes with perforations in the outer
tube graded with openings largest at center and smallest at ends,
all pointing downward.  This is illustrated in Fig. 5.  The slimes
were admitted to the outer tube and the CGAs were admitted
through the inner tube that had perforations in all directions
round the surface.  It must be remembered that the CGA bubbles
are of the order of 25 μm in diameter so that only small perforations were necessary.  The CGA met the slimes in the outer
tube and left through the perforations in the undersurface,
carrying the buoyant material upward while the sediment just
fell to the bottom.  It was found that as the froth passed the
baffle, there was a minor disturbance, and some froth reentered
the raffinate but was still buoyant enough to return to the surface.  The introduction of a second baffle a short distance further along cured this trouble.  In this way it was found possible
to get a total phosphorus recovery of 43%.

   The encouraging results with the phosphate slimes suggests
that the technique might have wider possibilities in the minerals
beneficiation areas.  Because of the rapid depletion of the higher
grade ores of minerals, it is inevitable that in the future attention

will have to be given to recovery from poorer ores. This will
necessitate crushing to finer sizes in order to liberate the desired
minerals. Because of this the material to be treated will be so
fine that conventional flotation will no longer be suitable. It is
possible that bubble-entrained floc flotation will enable such
fines to be treated, as particles of colloidal size can be floated.
It would be necessary to find a selective flocculant unique to
each case that would promote a floc surrounding the bubbles
but leave the gangue unfloatable. In such cases it may become
necessary to dilute the slurry very considerably. This should
not be a deterrant as the flotation is usually so complete that
the water could be recycled many times.

Another application of bubble-entrained floc flotation that
has had some measure of success but unfortunately has not been
suitably published is in the area of effluent water clean-up.
The method was tested on the effluent from a laundry that
specialized in cleaning industrial work clothes. The effluent
is particularly repulsive, consisting of black, murky water that
contained lint, soot, oil, and finely divided metallic particles.
Attempts to float these directly with CGA were unsuccessful.
If, however, some aluminum sulfate solution was added, the pH
adjusted, and an anionic CGA introduced, the aluminum hydroxide
or aluminum soap flocculated round the bubbles and at the same
time entrained all the solid particles, irrespective of their nature,
leaving a relatively clear water. A second flotation produced a
clear white water. The cost of such a process would be negligible when compared with the advantage of solving a serious
environmental problem. It would seem that there are many industrial plants where such clean-up would be desirable. There
are the black waters from the coal-washing plants; the red waters
from the bauxite extraction in the aluminum production industry;
the white effluent in the paper industry; and the complex effluents in the food-processing industries. It is not unlikely as
this technique becomes better known that authorities will insist
on such clean-up before the effluent is discharged into the
sewage system.

Another example of bubble-entrained floc flotation is the
flotation of algae and other microorganisms (15). As microorganisms are negatively charged, CGAs made from cationic surfactants can float these very effectively and completely. It is
believed that the mechanism is that of flocculation of the microorganisms by the cationic gas aphrons followed by flotation (see
also Ref. 16).

## IX. SEPARATION OF OIL FROM SAND

Aphrons can be used for separation of oils from solids such as
sand. This is somewhat akin to washing the sand, and CGAs,
which have excellent performance in detergency, can be used
for this purpose. For example, CGA could be applied in clean-
up after oil or gasoline spills, or where gasoline has leaked
underground from faulty storage tanks. Some preliminary studies
by Michelsen et al. (17) showed the feasibility of the method.
Bench studies have shown that an advancing front of CGA made
of an anionic surfactant percolating through a sand bed carries with
it most of the oil. The CGA must not be cationic because such a
CGA will not move through the sand bed but becomes blocked.
This, presumably, is because the cationic surfactant is attracted
to the negatively charged sand and makes it hydrophobic.

The CGA bubbles are small enough to enter channels greater
than 25 μm in diameter but some channels are less than that.
Until such time that CGA smaller than 25 μm can be made, which
seems unlikely, there will always be some limitation to this method
for scavenging oil underground. Nevertheless, as some of the
oil is removed it may be useful for partial clean-up. Some hazardous oils other than hydrocarbons could also be removed in
this way. The procedure would be to block the upper layers by
percolating through a cationic surfactant and then introducing
a CGA through a borehole underground and collecting the loaded
CGA at boreholes some distance away. Because of the blocked
"ceiling" the CGA has to move horizontally toward the wells.
Experiments in a transparent walled cell (2.13 m × 2.13 m × 12.7
cm) filled with medium-grade sand indicates that the CGA bubbles
do not collapse under such conditions.

Polyaphrons made of kerosene show promise in separating
the bitumen from tar sands. In tar sands, the bitumen is extremely viscous but flows readily when dissolved in kerosene.
When kerosene is in the form of aphrons, these seem to have the
ability to get between the bitumen and the sand, thus effectively
rolling up the bitumen. "Rolling up" is a term used in detergency
to describe the removal of an oil from a surface. In this process
the water-soluble surfactant, which stabilizes the soap film of
the aphron, adsorbs on the surface of the sand making it water-
wettable so that the oil cannot return to the sand. Double
trituration with polyaphrons leaves a perfectly clean sand.

## X.  PREDISPERSED SOLVENT EXTRACTION

Predispersed solvent extraction is a new process for separating solutes from aqueous solution by solvent extraction in which surfactants play an important part.  It has shown promise for extraction from extremely dilute solution very efficiently and quickly.  There are two unusual factors that contribute to this:

1.  There is no need for a mixing-settling stage, and
2.  The ratio of extracting solvent to pregnant solution can be very low, of the order of 1:1000 or even lower.

Mixer-settlers as used in industrial solvent extraction can be expensive, and for very dilute pregnant solutions the power costs can be significantly high and wasteful.  The reason for this operation is the necessity for comminuting the phases in order to achieve a large interface between the pregnant solution and the solvent in order to facilitate mass transfer.  But there is really no need to comminute both phases as transport occurs across the interface.  In fact, all that is necessary is to comminute one phase and clearly that should be the minor phase, namely, the solvent.  This can be achieved in an alternative way by conversion of the solvent into a polyaphron prior to the extraction.

As water is the continuous phase, addition of the polyaphron to water breaks it down into individual aphrons dispersed in the water.  In practice, a 50 fold dilution has proved to be suitable prior to addition to the pregnant solution.  Conversion into aphrons produces an enormous increase in interfacial area.  For example, 1 liter of a polyaphron of PVR 9 composed of aphrons of average diameter 2 μm would have an interfacial area for the solvent of 2700 $m^2$.  This is the reason the mixing-settling stage can be eliminated.

Although the solvent is usually lighter than water and would be expected to rise to the surface under the influence of gravity, this is usually very slow because of the small size of the aphrons.  Therefore, an essential feature of the new technique is the buoying of the aphrons to the surface by a flotation process using CGA.  For the same reason that oil droplets adhere to CGA bubbles, the microscopic liquid core aphrons adhere to the CGA bubbles which, being considerably larger than the solvent

aphrons, have correspondingly more buoyancy. Thus they rise
to the surface very quickly, carrying the solvent with them.
An essential step in this solvent extraction procedure is that
after the diluted polyaphron has been added to the pregnant
solution, an appropriate quantity of CGA is pumped into the cell
and left to stand for a few minutes, after which time the solvent
will be found at the surface having stripped the pregnant solution of the solute to be removed.

Predispersed solvent extraction can be achieved with very
simple equipment, namely, a column that has a capacity about
double the volume of the pregnant solution; entry points for
the pregnant solution, the diluted polyaphrons, and the CGA;
and a stopcock at the bottom of the column so that the raffinate
can be run off and separated from the loaded solvent. A sketch
of a laboratory cell is shown in Fig. 6.

A continuous cell has also been tested. This is simply a
trough, not unlike the one used for bubble-entrained floc flotation, with a weir at the end distal from the entry point at which
the solvent can overflow. Trial runs extracting copper from a
dilute solution in a continuous process has given 98% extraction,
and there is no reason why this could not be improved.

There is a very important distinction between the theory of
conventional solvent extraction and that of predispersed solvent
extraction. Both depend, of course, on a favorable distribution
coefficient of the solute between solvent and water. This will
be referred to as the extraction coefficient E. However, in a
mixing-settling extraction, transfer from the solution to the
solvent will cease as soon as the equilibrium is established, so
that there is a limit to the amount of extraction at each stage.
On the other hand, in predispersed solvent extraction the equilibrium is between each aphron and the pregnant solution at the
point when the aphron is just about to leave the solution and
enter the surface, effectively removing it from the equilibrium.
Each aphron therefore leaves a portion of the solution that is
slightly depleted in solute. Clearly, each aphron cannot be in
equilibrium with the total volume of pregnant solution, so it is
convenient to define a depletion volume, which is the volume of
pregnant solution with which the aphron has time to establish
equilibrium just prior to leaving it, as sketched in Fig. 7.

The depletion volume would, of course, be dependent on
factors such as temperature, diffusion coefficient, and concentration, and would be an extremely small volume. However,
there are an enormous number of aphrons all following one
another, with each extracting, irreversibly, a small amount of

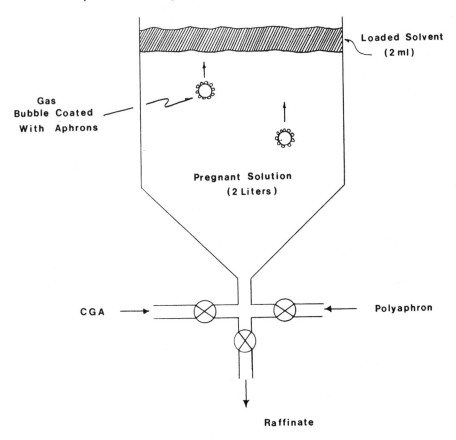

FIG. 6   Predispersed solvent extraction cell.

solute.   The net effect is that extraction in this way can be
virtually complete.   A useful analogy is the analytical procedure
for washing a precipitate, where it is more efficient to wash a
few times with the wash liquid divided into small volumes than
to extract once with the full amount.

   If a is the original number of moles of solute and $x_n$ is
number of moles remaining after extraction by n aphrons seriatim,
it can be shown (18) that

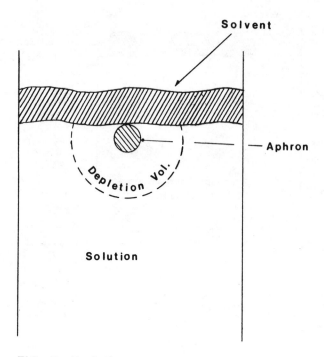

FIG. 7   Depletion volume.

$$\frac{x_n}{a} = p^n$$

where $p = V_D/(Ev + V_D)$, $V_D$ is depletion volume, E is
extraction coefficient, and v is volume of each aphron. It is
the exponent n that makes this form of solvent extraction so
efficient. If f is the ratio of total volume of solvent to total
volume of pregnant, then $n = fV_D/v$. If the aphron is delayed
at interface for 0.5 sec and diffusivity is $10^{-5}$ cm$^2$/sec, then
$V_D$ is approximately 400 μm$^3$. Assuming f = 0.3 and v = 1 μm$^3$,
then n = 120. Then, if E = 10, $x_n/a$ = 0.052, whereas for a
single conventional extraction, $x_n/a$ would be 0.25. On the
other hand, if E = 100, $x_n/a$ would be $2.3 \times 10^{-12}$, whereas for
a single conventional extraction it would be 0.032.

There are two factors that determine the efficiency of the
extraction. One is the value of p, which must be as small

a fraction as possible. Thus the larger the value of E, the extraction coefficient, the smaller p will be. Second, the value of n, the exponent, must be as large as possible. This would be determined by the ratio of volume of extracting solvent to the volume of the pregnant solution f and by the size of the aphrons. The smaller the aphrons, the larger n will be. Since the aphrons are removed by flotation, a reduction in the size of the aphrons does not pose a problem unless they are made so small that they are virtually solubilized. It must be made clear that the above equations have not yet been tested quantitatively.

The danger of possible foaming can be avoided by a judicious choice of surfactant for the CGA, taking advantage of the contrafoam phenomenon as reported by Sebba (19). This is the fact that a foam can often be broken by contacting it with a foam of opposite charge. If, for example, the polyaphron has been made using an anionic surfactant for the encapsulating soapy film, then the CGA that is to be used in the flotation stage should be made with a cationic surfactant. In this way the amount of foam is minimal provided no more CGA is used than is necessary for complete flotation.

As an example of the power of this technique, an oil-soluble dye, waxoline blue, as supplied by ICI, was dispersed in water by dissolving it in alcohol and then adding it to 2 liters of water to produce a concentration of 2 ppb. At the concentration the dye behaved as though it were dissolved. The solution was placed in the cell, and 1 ml of a kerosene polyaphron of PVR 20 was diluted 50-fold in water and then added to the pregnant solution. The polyaphron had been made using a solution of sodium dodecyl benzenesulfonate at a concentration of 4 g/liter. As this was an anionic surfactant, the CGA was made using the cationic surfactant dodecyltrimethylammonium chloride at a concentration of 0.5 g/liter. About 200 ml of CGA was introduced slowly into the pregnant solution, and within about a minute a layer of kerosene, colored by the dye, collected above the solution. In this case, of course, the extraction coefficient must have been very high as the dye was an oil-soluble dye. Nevertheless, the speed and effectiveness of extraction at that low concentration is very encouraging.

There is no reason why extraction could not be effected from even more dilute solutions in this way. It would seen that this method could be developed for extraction of hazardous organic solutes from aqueous effluents, even if very dilute, before discharge into streams.

Copper has been extracted from a solution of copper sulfate
($10^{-4}$ M), using a 10% solution of a liquid ion exchanger,
LIX-64N, an α-hydroxyoxime, in kerosene made up as a polyaphron
with sodium dodecyl benzene sulfonate as the surfactant. In
this case again the CGA was cationic, so that there was not any
appreciable foam. The extraction in one pass was so effective
that copper was undetectable in the raffinate. The procedural
details are given in Ref. 20. Uranium, as the complexed anion,
has been concentrated from a solution of uranyl acetate (5 × $10^{-5}$
M) using a polyaphron made up of 10% dodecylamine in kerosene,
the surfactant in this case being cationic dodecyltrimethylammonium
chloride. For that reason the CGA was made anionic, the surfactant being sodium dodecyl benzene sulfonate (0.3 g/liter).
Uranium at a similar concentration has also been extracted using
a polyaphron made of tributyl phosphate dissolved in kerosene.
Chromate at a concentration of 0.01 g/liter has been completely
floated using 10% tricaprylamine dissolved in kerosene as collector.

There have been cases where some residual turbidity in the
raffinate has been observed. This is probably due to some
solubilization of the oil phase, although it does not seem to diminish the efficiency of the extraction. Should this be an undesirable feature, it can easily be remedied by adding a very
small quantity of aluminum sulfate solution to the raffinate to
produce a concentration of about 0.0001 M, adjusting the pH to
5.5 and then introducing about 100 ml of a CGA made of sodium
dodecyl benzene sulfonate. An aluminum soap quickly floats to
the surface containing all the suspended material, leaving a
perfectly clear solution.

The predispersed solvent extraction technique shows promise,
not only for removal of traces of undesirable materials, organic
as well as inorganic, from aqueous solution, but it might also
be used as a preliminary concentration technique in analytical
chemistry because it is so quick and simple.

# REFERENCES

1.  A. W. Adamson, *Physical Chemistry of Surfaces*, 4th ed.,
    Wiley, New York, 1982, p. 478.
2.  F. Sebba, *Chem. Ind.*. 367 (1984).
3.  F. Sebba, *Chem. Ind.*, 91 (1985).
4.  F. Sebba, Report OWRT/RU-82/10 to U.S. Department of
    Interior, 13 (1982).

5.  W. Drost-Hansen, *J. Colloid Interf. Sci.*, *58*: 251 (1977).
6.  F. Sebba, *Nature*, *184*: 1062 (1959).
7.  F. Sebba, *Ion Flotation*, Elsevier, Amsterdam, 1962.
8.  R. B. Grieves, in *Treatise on Analytical Chemistry, Part 1, Theory and Practice* Vol. 5, 2nd ed. (P. J. Elving and I. M. Kolthoff, eds.), Wiley, New York, 1982, p. 371.
9.  S. Ciriello, S. M. Barnett, and F. J. Deluise, *Sep. Sci. Technol.*, *17*: 521 (1982).
10. P. I. Shea and S. M. Barnett, *Sep. Sci. Technol.*, *14*: 757 (1979).
11. S. M. Barnett and S. F. Lin, *Proc. Conf. Seafood Waste Management* (W. S. Otwell, ed.), Florida Sea Grant College, University of Florida, Orlando, 1981, p. 176.
12. S. M. Barnett, private communication.
13. W. Auten and F. Sebba in *Solid–Liquid Separation* (J. Gregory, ed.), Ellis Horwood Ltd., Chichester, Sussex, England, 1984, p. 41.
14. S. Alexander, A. de Moor, and F. Sebba, *J. Sep. Process Technol.* (in press).
15. S. S. Honeycutt, D. A. Wallis, and F. Sebba, *Biotechnol. Bioeng. Symp.*, *13* (1983).
16. D. A. Wallis, D. L. Michelsen, F. Sebba, J. K. Carpenter, and D. Houle, *Biotechnol. Bioeng. Symp.*, *15*: 399 (1983).
17. D. L. Michelsen, D. A. Wallis, and F. Sebba, *Proc. 3rd Int. Congr. Chem. Eng.*, Society of Chemical Engineers of Japan, Tokyo, Vol. 3, 1986, p. 592.
18. F. Sebba, *Sep. Purif. Methods*, *14*: 127 (1985).
19. F. Sebba, *Nature*, *197*: 1195 (1963).
20. A. J. Aggarwal, A. Rodarte, and F. Sebba, *J. Sep. Process Technol.* (in press).

# 5

# Microemulsion-Based Separations

STIG E. FRIBERG* and PARTHASAKHA NEOGI    Departments of
Chemistry and Chemical Engineering, University of Missouri-Rolla,
Rolla, Missouri

*Present affiliation: Department of Chemistry, Clarkson College of
Technology, Potsdam, New York.
This work was supported by DOE Grant No. DOE-AC02 83 ER 13083.

## SYNOPSIS

Microemulsions as extraction vehicles offer several advantages in reduced toxicity and low energy requirements for the total process due
to the reversibility of the process. The phases involved have low
energies of stabilization and the entire cycle of solubilization and break-
up of microemulsions is contained in a temperature range of about 20°C.
     However, the complexity of microemulsion systems may lead to
the appearance of new structures during the extraction process due
to strong variation in diffusion rates for different components. In
particular, liquid crystals appear, even if they are absent on the
pseudoternary diagrams. Two cases in which temporary liquids
crystals strongly influence extraction processes are presented. The
observations span over tens to hundreds of days since the growth
and disappearance of the intermediate phases are slow. A brief
description of the efforts to quantitize these effects using the fundamental transport theory is also included.

## I. INTRODUCTION

Extraction processes using surfactant aggregates offer several advantages in comparison to traditional extraction processes using
organic solvents. First, environmental concerns favor the absence
of organic solvents in any process and second, the small energies involved in structural changes of colloidal association aggregates make
the separation process appealing from an economic point of view (1).
     The last 20 years has seen the behavior of surfactants in water-
oil systems applied in a multifold of new extraction areas reaching
from the double emulsion (liquid membranes) (2) over to microemulsions in tertiary oil recovery (3-5) the ingenious micellar-enhanced ultrafiltration (Chapters 1 and 2 of this book), to applications
in biotechnology (Chapter 3 of this book).
     The tertiary oil recovery process (3-5) focused the interest on
the applications of microemulsions (6,7) in extraction processes.
Microemulsions are related to micellar solutions (8) and a short review of introductory investigations into the simplest extraction, those
of an inclusion of an organic material into an aqueous micellar solution, is of value.
     Evans and collaborators (9) studied the mass transfer of a solid
fatty acid from a spinning disk into an aqueous micellar solution.

Their results were interpreted as the limiting factor being the
adsorption of micelles at the interface. The slow stage of the
process included the desorption of the micelle from the solid surface.
Carroll (10) and Caroll et al. (11) investigated the rates of solubilization of a liquid hydrocarbon into an aqueous micellar solution.
These authors also interpreted their results as indicating the
processes with the micelle at the interface to be the determining
factors. They could show that the diffusion of hydrocarbon molecules through the water to the micelles was without significance for
the process with an aliphatic hydrocarbon. Their detailed interpretation revealed the restructuring of the micelle during the solubilization to be the rate-determining step.

A somewhat similar mechanism was proposed by Tondre and
Zana (12) for the uptake of oil from emulsion droplets into a microemulsion under rapid mixing. They found the dissolution rates
greater than values proportional to the reciprocal time of mixing
and concluded that the diffusion-controlled collisions between emulsion and microemulsion droplets were the rate-determining step.

The following development in the use of microemulsions for the
tertiary oil recovery process (3—5) has been reviewed by Miller
and Qutubuddin (13). Neogi (14) analyzed the special problems
related to the statics and dynamics of microemulsions flooding of
porous reservoirs. Neustadter (15) reviewed the surfactant-co-
surfactant systems useful for oil field applications.

With these recent review articles available a publication of
general nature in the area of colloidal extraction would, at present,
appear redundant. With this in mind, we have chosen to describe
some special problems encountered in the nonstirred extraction
process that had hitherto not been observed. They do not exist
in the stirred processes due to the fact that the mixing eliminates
the concentration gradients, which appear during diffusion processes
with extended pathlengths. A contact zone between pure water
and a microemulsion may, hence, depending on the relative diffusion
rates, display a series of different structures under nonstirring
conditions.

Hence, in this chapter we will focus on the significance of
association structures other than micelles for the extraction process
as exemplified in two systems. In both systems a temporary liquid
crystalline phase appears: its presence is a liability in the first
system of nonionic surfactants and may be an advantage in the
second one.

## II. AN EXTRACTION/SEPARATION METHOD USING THE TEMPER-
## ATURE DEPENDENCE IN A NONIONIC SURFACTANT SYSTEM

Nonionic surfactants display extreme temperature dependence of
their solubilization capacity of hydrocarbons as observed more than
20 years ago by Shinoda (16). A detailed description of these phase
equilibria has recently been given (17) but the simplified form of the
behavior according to Fig. 1 (a)–(c) is sufficient for the present
purpose.

The essential information for extraction purposes is the difference
between the solubilization capacity of the nonionic surfactants below
the HLB temperature [Fig. 1(a)], at that temperature level [Fig. 1(b)]
and above it [Fig. 1(c)]. At the low temperature the surfactant is

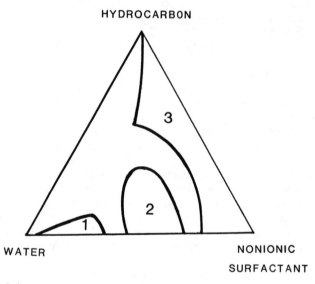

HYDROCARBON

WATER                                                NONIONIC
                                                    SURFACTANT

(a)

FIG. 1 The three phases in a system of water and nonionic surfactant
(polyethylene glycol alkyl ether) and hydrocarbon. (a) Low temperature: 1, normal micellar solution; 2, lamellar liquid crystal; 3,
hydrocarbon surfactant solution in solubilized water. (b) HLB temperature: 1, surfactant phase; 2, lamellar liquid crystal; 3, hydrocarbon surfactant solution in solubilized water. (c) High temperature:
2, lamellar liquid crystal; 3, hydrocarbon surfactant solution in
solubilized water.

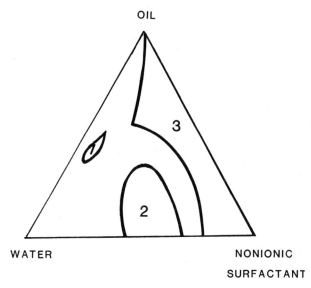

OIL

WATER          NONIONIC
               SURFACTANT

(b)

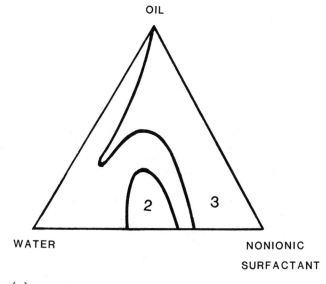

OIL

WATER                    NONIONIC
                         SURFACTANT

(c)

FIG. 1  (Continued)

soluble in water and normal micelles are formed; at high temperatures, on the other hand, the solubility is limited to the hydrocarbon. At the HLB temperature [Fig. 1(b)], an isotropic solution is formed containing approximately equal amounts of water and hydrocarbon [18] with very low surfactant concentration (~6-8%).

These changes are in fact a temperature-dependent inversion, a subject that has attracted considerable attention from theoreticians. Inversion as such has been treated in a theoretical model by Widom (19). As for the structure and stability of this middle phase as surfactant phase, there are several factors to be evaluated. The first condition *sine qua non* is an ultralow interfacial tension as pointed out early by Ruckenstein and Chi (20). For the middle phase, the observations by Miller et al. (21) are relevant in pointing out that ultralow interfacial tensions require a surface zone of a thickness comparable to the radius of the droplet. With an ultralow interfacial tension, the surface free energy contribution to the total free energy may be overcome by the dispersion entropy of the drop-lets and thermodynamic stability is achieved (20). For the surfactant phase, a description of the structure in the form of spherical droplets is less than adequate and a relation between the dispersion entropy contribution and the random arrangement of the oil-water interface is a more appropriate approach (22).

Whatever the state of theoretical understanding of the structure and stability of these systems, it is obvious that the phase diagrams in Fig. 1(a)-(c) in principle offer an elegant and efficient extraction/separation method with extremely small energy requirements (1).

## III.  A LOW-ENERGY EXTRACTION/SEPARATION METHOD

The method is based on the fact that the difference in temperature between the conditions in Fig. 1(a) and (b) is only 15-20°C. Hence, a complete extraction/separation process is envisioned in the following manner with the only requirements of heating the system by approximately 20°C.

An aqueous solution with X% surfactant Fig. 2(a) is brought into contact with an oil-carrying substrate at the elevated temperature to give the phases of Fig. 2(b). Extraction now takes place spontaneously.

Separation is obtained by reduction of the temperature to the conditions in Fig. 2(a). The separation results in an aqueous phase with most of the surfactant plus some solubilized oil at point P and an oil phase with some surfactant and solubilized water at point Q Fig. 2(c).

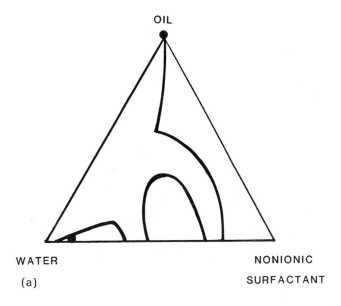

WATER                          NONIONIC

(a)                           SURFACTANT

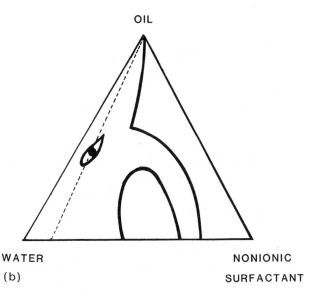

WATER                          NONIONIC

(b)                           SURFACTANT

FIG. 2  The composition marked in (a) is brought into contact with oil and the temperature is raised to the HLB temperature (b). Extraction now takes place and by reducing its temperature to (a) an oil phase of composition Q is separated in equilibrium with an aqueous solution of composition P.

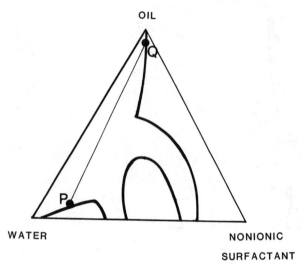

(c)

FIG. 2    (Continued)


        The composition of point Q is essential for the process because the
surfactant in the oil phase is lost in the process and its cost should
be balanced against the reduction in the expenses for separation by
other means.  In a preliminary experiment using a well-defined nonionic surfactant, the water content in oil was 0.02% and the surfactant
content 0.37%.  This latter value is not insignificantly low, but at the
time it was felt to be at a level that was encouraging for further
investigations.
        These gave a surprising result.  The problems encountered with
the method are not found in the separation stage but in the extraction
state.  When the aqueous solution as in Fig. 2(a) is contacted with
pure oil, the formation of the isotropic liquid surfactant phase is not
direct.  Instead the diffusion of different components in different
directions cause other phases to appear.  A typical case will first
be described, followed by an analysis of the diffusion pathways that
cause the formation of the different phases.

## IV. EXTRACTION EXPERIMENT: OVERALL APPEARANCE

The appearance in a test tube after the hydrocarbon had been gently poured onto the aqueous solution of the nonionic surfactant to give a total composition in the surfactant phase [Fig. 2(a)] at equilibrium is given in Fig. 3.

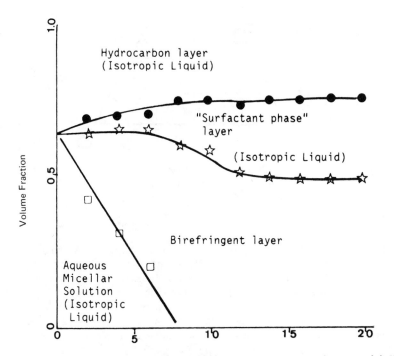

FIG. 3 Relative variation of different layers vs. time. Initially, 1.169 g of n-decane was placed on top of 3.00 g of aqueous micellar solution 74.92% water and 25.08% surfactant. The initial height of the hydrocarbon layer was 1.54 cm.

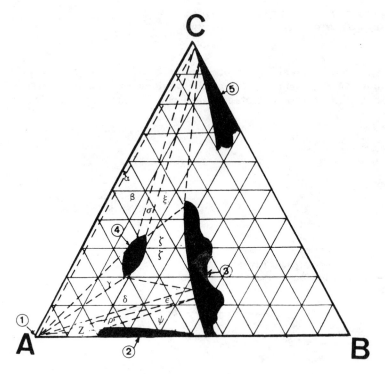

FIG. 4 The system with tielines marking two- and three-phase areas (see text). A, B, C are water, surfactant, and oil.

Two features are immediately evident from the figure. At first the amount of oil removed is fairly small; about one-third of the available amount. The aqueous layer disappears with constant rate to be entirely depleted at 750 hr. By that time a birefringent layer occupies the initial volume of the water. From that time the volume of the surfactant phase increases; the birefringent layer is reduced correspondingly and the volume of the oil phase remains approximately constant. These results may be analyzed first when the equilibrium conditions between the phases involved has been clarified.

The part of the phase diagram that is relevant to the temperature of extraction is presented in Fig. 4. The extraction process is influenced by five phases:

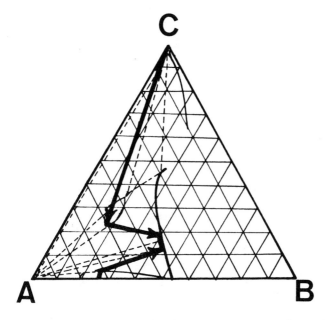

FIG. 5 The arrows mark the transport path from the micellar solution
and n-decane. A, B, C are water, surfactant, and oil.

1. Water
2. Micellar solution
3. Lamellar liquid crystals
4. Surfactant phase
5. Oil solution

The equilibria are best characterized by observing the water phase.
Counted clockwise it is in a two-phase equilibrium with the oil phase
($\alpha$), a three-phase equilibrium with the oil phase and the surfactant
phase ($\beta$), a two-phase equilibrium with the surfactant phase ($\gamma$), a
three-phase equilibrium with the surfactant phase and the lamellar
liquid crystal ($\delta$), a two-phase equilibrium with the lamellar liquid
($\epsilon$), a three-phase equilibrium with the lamellar liquid crystal and the
aqueous micellar solution ($\rho$), a two-phase equilibrium with the latter
($\Psi$) and between water and the micelle solution ($\iota$). In addition, there
are a two-phase equilibrium between the surfactant phase and the
lamellar liquid crystal ($\zeta$): a three-phase equilibrium between the
surfactant phase, the lamellar liquid crystal, and the oil phase ($\xi$);

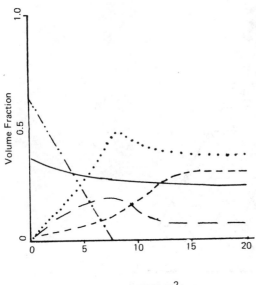

TIME $10^2$ hours

FIG. 6  The relative volume of all involved phases was calculated at
a function of time:  (—) n-decane, (- -) lamellar liquid crystal,
(--) surfactant phase, (- ·· -) aqueous micellar solution, (...) water.

and a two-phase equilibrium between the surfactant phase and the
oil phase ($\sigma$).

A comparison of Figs. 3 and 4 indicates a transport of hydrocarbon
from its phase via the surfactant phase to the lamellar liquid crystals,
whose demand for surfactant would be filled by a corresponding transport from the aqueous micellar solution as indicated in Fig. 5.

These pathways are correct in principle, but an estimation of the
mass balance shows that the birefringent layer cannot be a single-
phase liquid crystalline structure.  Taking the surfactant needed for
the surfactant phase and the phase equilibria (Fig. 4) into consideration, it becomes obvious that the birefringent layer is a two-phase
system of water and lamellar liquid crystal ($\varepsilon$, Fig. 4).

A calculation of the amount in the different phases is shown in
Fig. 6.  It is interesting to note the growth of the water and the
liquid crystal phase during the initial reduction of the micellar solution and their reduction after the micellar solution was depleted.
After that time the surfactant phase began to increase.

The results also made possible an analysis of the rate-determining steps in the transport chain (8).

## V. ANALYSIS OF THE INDIVIDUAL TRANSPORT STEPS

Four layers are formed in the extraction process. They are marked from bottom to top (Fig. 7). They are marked I (micellar), II (birefringent), III (surfactant), and IV (hydrocarbon). The three interfaces are marked $\alpha$ (I/II), $\beta$ (II/III), and $\gamma$ (III/IV). The bottom part of each layer is marked by a plus sign and the top with a minus sign. Hence, the interface between the birefringent phase and the surfactant phase implies the contact $\beta_-/\beta_+$.

A complete analysis of the factors involved in the transport would of necessity be too complex to be meaningful and the discussion is limited to a few pertient factors.

The oil phase in contact with the surfactant phase contains but small amounts of water, which is molecularly dispersed at these low concentrations. The diffusion coefficients for molecularly dispersed water and surfactants are at least a magnitude higher than those in

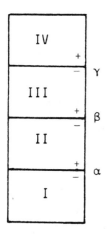

FIG. 7 The arrangement of the four phases: micellar phase (I), birefringent phase (II), surfactant phase (III), and oil (IV); the three interfaces are labeled from below as $\alpha$, $\beta$, $\gamma$.

the surfactant phase and the liquid crystal, and it is justified to consider the diffusion in the oil phase not to be a rate-determining step.

The depletion of the aqueous micellar solution took place linearly with time. If the process were diffusion-dependent, a dependence on the square root of time would have been found. This was not the case and it is reasonable to consider diffusion in the aqueous micellar solution as without influence on the total transport.

Hence, the material balance for the aqueous micellar solution is simply described.

$$\rho^{I} \frac{dl^{I}}{dt} = -R \left( 1 + \frac{w_{\alpha-}}{s_{\alpha-}} \right) \tag{1}$$

in which R is the rate (mass/time interfacial area) at which the surfactant leaves phase I.

Analysis of the transport rates gave the material balance for the birefringent layer as

$$\rho^{II} \frac{dl^{II}}{dt} = R \left[ 1 + \frac{w_{\alpha-}}{s_{\alpha-}} \right] + \frac{\rho^{III} D_{o}^{III}}{l^{III}} ({}^{o}\gamma_{-} - {}^{o}\beta_{+}) - Q \left[ \frac{{}^{o}\beta_{+}}{s_{\beta_{+}}} - \frac{{}^{o}\beta_{-}}{s_{\beta_{-}}} \right]$$

$$- Q \left[ 1 + \frac{{}^{o}\beta_{-}}{s_{\beta_{-}}} \right] \tag{2}$$

in which Q is the corresponding uptake of surfactant from the lamellar liquid crystal to the surfactant phase.

Finally, the material balance of the surfactant phase:

$$\rho^{III} \frac{dl^{III}}{dt} = \frac{Q}{s_{\beta_{+}}} + \frac{\rho_{LC}^{II} D_{w}^{II}}{l^{II}} ({}^{w}\alpha_{+} - {}^{w}\beta_{-}) - Q \left[ \frac{w_{\beta_{+}}}{s_{\beta_{+}}} - \frac{w_{\beta_{-}}}{s_{\beta_{-}}} \right] \tag{3}$$

w, s, and o are weight fractions of water, surfactant, and oil and the subscripts determine their locations. The equations fit the change of layer thickness l, as labeled in Fig. 7, and gave the following values:

Adsorption rate at $\alpha$:  $R = 7.78 \times 10^{-4}$ g/(cm$^2$ · hr)
Adsorption rate at $\beta$:  $Q = 1.7 \times 10^{-4}$ g/(cm$^2$ · hr), and
Diffusivities:  $D_o^{III} = 2.4 \times 10^{-7}$ cm$^2$/sec

$$D_o^{II} = 1.3 \times 10^{-6} \text{ cm}^2/\text{sec}$$

The value for the transport of surfactant from the liquid crystal to the surfactant phase agrees with the value of palmitic acid into an aqueous solution micellar solution of sodium dodecyl sulfate $1-2 \times 10^{-4}$ M (9). The transport from the aqueous micellar solution was obviously considerably higher. The diffusion coefficients for water and oil are also of the same magnitude as those reported elsewhere.

The results reveal the fact that what appears to be a simple extraction process using the microemulsion concept may become complicated due to variation in transport rates. This variation may also give rise to temporary phases, which are not found at equilibrium.

For such a use the temporary liquid crystalline phase should be formed at the interface between the microemulsions and water; the liquid crystal formation between the oil phase and the microemulsion is obviously of no use as a viscosity enhancer. A preliminary investigation (23) demonstrated the feasibility of these temporary liquid crystals.

## VI.  TEMPORARY LIQUID CRYSTALS IN MICROEMULSIONS STABILIZED BY AN IONIC SURFACTANT/ALCOHOL COSURFACTANT COMBINATION

A combination of a microemulsion with low water content (P, Fig. 8) and pure water (Q, Fig. 8) in the correct amounts to give a microemulsion with high water content (R, Fig. 8) will do so immediately at gentle agitation. However, if the microemulsion is gently layered on top of the water, the homogenization by diffusion will take a long time (23).

The important result is that the relative diffusivities of the components may lead to compositions in the two phases that deviate from the straight line between P and Q (Fig. 8).

Such is the case in a system of water, sodium dodecyl sulfate, and pentanol/decane (1/2 weight ratio) (Fig. 9). The appropriate amounts of water were combined with microemulsions of compositions P, Q, and R to form microemulsions close to the left limit in the solubility region in the top part of the diagram.

The behavior of the samples is different for different cosurfactant/surfactant ratios. A sample at point P, Fig. 9 shows changes in the layers according to Fig. 10(a). In the initial period ($\sim 10$

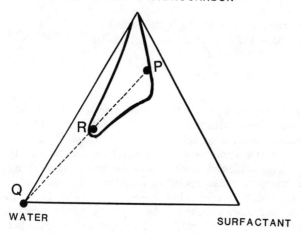

COSURFACTANT & HYDROCARBON

P

R

Q

WATER                                    SURFACTANT

FIG. 8  A combination of a microemulsion and minimum water content (P) and water (Q) will form a microemulsion with high water content (R).

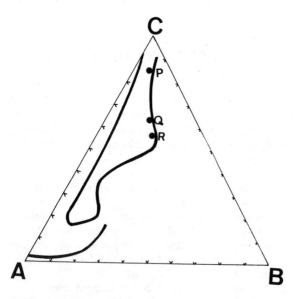

C

P

Q

R

A                                          B

FIG. 9  Microemulsions with compositions P, Q. R were gently layered on top of pure water in amounts to form microemulsions with maximum water content.  A, B, C are water, surfactant, and oil.

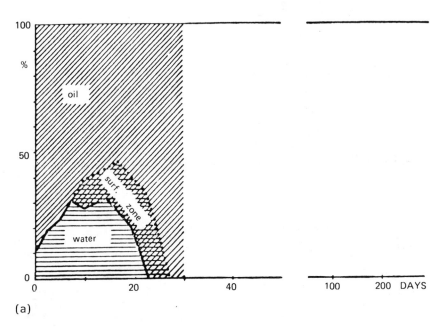

(a)

FIG. 10   The microemulsions and compositions P, Q, R (Fig. 9),
showed different behavior when contacted with water (a).   The P
microemulsion initially accepted microemulsion components into the
water but after 10 days the directional transport was reversed (b).
The microemulsion Q (Fig. 9) showed a similar behavior up to 20
days, but after that time the direction of transport was reversed
once more.   (c) The microemulsion composition R forms a temporary
lamellar crystal layer between the microemulsion and the water.

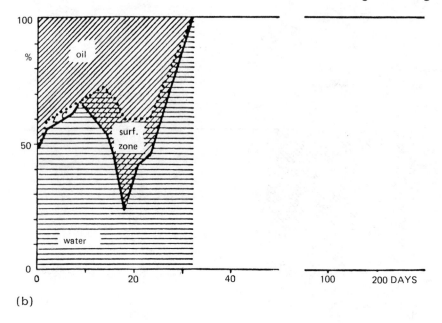

(b)

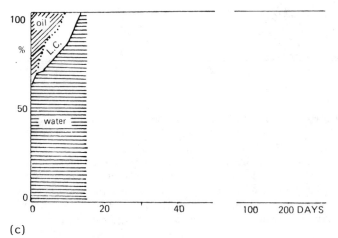

(c)

FIG. 10   (Continued)

days) the diffusion from the oil phase is predominant; after that time transport in the opposite direction is the governing factor. The latter part is characterized by a thick, $\sim 1/2$-cm, diffuse surface zone. The composition at Q shows the same pattern in Fig. 10(b) but now with further shift of direction. The composition at R finally gives rise to a liquid crystal between the oil and water [Fig. 10(c)].

The potential for such a temporary liquid crystal to serve as a viscosity enhancer in the microemulsion flooding is self-evident. It forms in the contact zone between pure water and a microemulsion leaving the hydrocarbon/microemulsion interface without a barrier. This means that the extraction of oil by the microemulsions is not hindered but "fingering" is reduced due to the high viscosity of the liquid crystal.

## REFERENCES

1.  T. D. Flaim and S. E. Friberg, *Sep. Sci. Technol.*, *16*: 1467 (1981).
2.  R. P. Cahn and N. N. Li, *J. Membrane Sci.*, *1*: 129 (1976).
3.  R. L. Reed and R. N. Healy, in *Improved Oil Recovery by Surfactant and Polymer Flooding* (D. O. Shah and R. S. Schechter, eds.), Academic Press, New York, 1977, p. 383.
4.  D. O. Shah, V. K. Vansal, K. S. Chan, and W. C. Hsieh, in *Improved Oil Recovery by Surfactant and Polymer Flooding* (D. O. Shah and R. S. Schechter, eds.), Academic Press, New York, 1977, p. 293.
5.  M. Baviere, W. H. Wade and R. S. Schechter, in *Surface Phenomena in Enhanced Oil Recovery* (D. O. Shah, ed.), Plenum Press, New York, 1981, p. 117.
6.  I. D. Robb, (ed.), *Microemulsions*, Plenum Press, New York, 1982.
7.  S. E. Friberg and R. I. Venable, in *Encyclopedia of Emulsion Technology* Vol. 1, (P. Becher, ed.), Marcel Dekker, New York, 1983, p. 287.
8.  P. Neogi, M. Kim, and S. E. Friberg, *Sep. Sci. Technol.*, *20*: 613 (1985).
9.  J. A. Schaeiwitz, F.-C. Chan, E. L. Cussler, and D. F. Evans, *J. Colloid Interf. Sci.*, *84*: 47 (1981).
10. B. J. Carroll, *J. Colloid Interf. Sci.*, *79*: 126 (1981).
11. B. J. Carroll, B. G. C. O'Rourke, and A. J. I. Ward, *J. Pharm. Pharmacol.*, *34*: 287 (1982).

138                                                    Friberg and Neogi

12. C. Tondre and R. Zana, *J. Dispersion Sci. Technol.*, *1*: 179 (1980).
13. C. A. Miller and S. Qutubuddin, in *Interfacial Phenomena in Apolar Media* (H. S. Eicke and G. D. Parfitt, eds.), Marcel Dekker, New York, 1987, p. 117.
14. P. Neogi, in *Microemulsions: Structure and Dynamics* (S. E. Friberg and P. Bothorel, eds.), CRC Press, Boca Raton, La., 1987, p. 197.
15. E. L. Neustadter, in *Surfactants* (Th. F. Tadros, ed.), Academic Press, London, 1984, p. 277.
16. H. Arai and K. Shinoda, *J. Colloid Interf. Sci.*, *25*: 396 (1967).
17. S. E. Friberg, in *Interfacial Phenomena in Apolar Media* (H.-F. Eicke and G. D. Parfitt, eds.), Marcel Dekker, New York, 1987, p. 93.
18. S. E. Friberg and I. Lapczynska, *Prog. Colloid Polym. Sci.*, *56*: 16 (1975).
19. B. Widom, *J. Chem. Phys.*, *81*, 1030 (1984): *J. Phys. Chem.*, *88*: 6508 (1984).
20. E. Ruckenstein and J. C. Chi, *J. Chem. Soc., Faraday Trans. II*, *71*: 1690 (1975).
21. C. A. Miller, R. Hwan, W. J. Benton, and T. Fort, *J. Colloid Interf. Sci.*, *61*: 554 (1977).
22. Y. Talmon and S. Prager, *J. Chem. Phys.*, *76*: 1535 (1982).
23. S. E. Friberg, M. Podzimek, and P. Neogi, *J. Dispersion Sci. Technol.*, *7*: 57 (1986).

# 6

# Liquid–Coacervate Extraction

NANCY D. GULLICKSON, JOHN F. SCAMEHORN, and
JEFFREY H. HARWELL* Institute for Applied Surfactant
Research, University of Oklahoma, Norman, Oklahoma

*Present affiliation:* Directorate for Engineering, Division of Chemistry, Biochemistry, and Thermal Engineering, National Science Foundation, Washington, D.C.

Financial support for this work was provided by Oklahoma Mining and Minerals Resources Research Institute and the OU Energy Research Institute. GAF Corp. donated the surfactants used.

## SYNOPSIS

When an aqueous solution containing a nonionic surfactant is heated above its cloud point, the solution may separate into two phases: a dilute phase and a phase containing the surfactant in high concentration, which is also called the coacervate phase. An organic solute originally present in solution will tend to become concentrated in the coacervate phase. In this study, separation of n-alcohols and 4-tert-butylphenol using this liquid-coacervate extraction technique was examined. The ratio of concentrations of the solute in the coacervate to that in the dilute phase could exceed 500 with phase volume ratios exceeding 20, indicating the high efficiency of this separation technique.

## I.  INTRODUCTION

As the temperature of an aqueous solution of nonionic surfactant is increased, a temperature may be reached where the solution turns cloudy; this temperature is referred to as the cloud point (1).  Above the cloud point, the solution may separate into a concentrated or coacervate phase, and a dilute phase.  In this case, the cloud point is a lower consolute solution temperature. The weight fraction of surfactant in the coacervate can be quite high, e.g., values exceeding 20% have been reported (2).  The concentration of surfactant in the dilute phase can be very low but is gnerally above the critical micelle concentration (CMC) (3-8).

The phase behavior of aqueous nonionic surfactant solutions can be very complex.  The cloud point phenomenon is only observed over a narrow region of the entire phase space (4).  For example, at a temperature above the cloud point, the multiphase solution may become a single isotropic phase; this is an upper consolute solution temperature.  Above the cloud point more than two phases can form, since the surfactant is usually polydisperse, without violation of the Gibbs phase rule; we have observed this behavior.  Since the cloud point is a critical point, the difference in concentration in the coacervate and dilute phases increases as the solution temperature increases above the cloud point (2), i.e., exactly at the cloud point, the two phases have identical compositions.  Finally, it is important to note that solutions may exhibit cloud points but not phase-separate (at least, not in a reasonable time scale).  The systems described here were chosen after the screening of numerous nonionic surfactants.

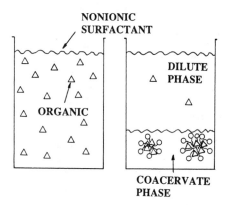

FIG. 1   Schematic of liquid-coacervate extraction.

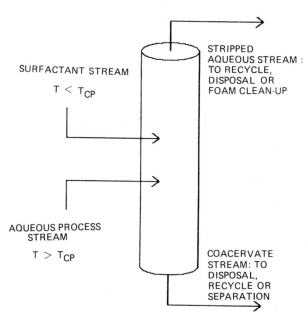

FIG. 2   Schematic of a continuous liquid-coacervate extraction.

If an organic solute is originally present in an aqueous stream and a nonionic surfactant (or concentrated surfactant solution) is added to the water, where the water temperature is above its cloud point, the organic solute will tend to concentrate itself into the coacervate phase. This liquid-coacervate extraction process is shown in Fig. 1. After equilibration, the two phases may be separated in a phase splitter, completing the separation. If the dilute phase contains solute and surfactant in low enough concentrations, it can be recycled to the process or emitted to the environment. The organic solute and/or the surfactant can be separated from the coacervate phase (e.g., by a foam separation) and sold, reused, or disposed. If multiple contacting units are necessary, the process can be run continuously and staged as shown in Fig. 2, as with traditional liquid-liquid extractions.

## II.   BACKGROUND ON TWO-PHASE AQUEOUS–AQUEOUS EXTRACTIONS

The addition of fairly dilute material to an aqueous solution, resulting in a phase separation, and utilization of this to extract solute preferentially into one phase, has received much attention. An important use of liquid-liquid extraction is in the purification and concentration of proteins and other biomolecules. In order to maintain the biological activity of these molecules, it is important that the biomolecules and proteins remain in an aqueous environment. A great amount of research has been conducted using two or more dissimilar water-soluble polymers in an aqueous solution to effect a phase separation. These phase systems have been shown to partition various biomolecules (9–14) and proteins (9,15–18).

The polymer concentrations in solution affect the interfacial tension between the two phases which form, which in turn has been shown to affect the partitioning of the biomolecules between the phases. Other factors affecting the partitioning of biomolecules and proteins are polymer molecular weight (10,16,18,19) and the molecular weight of the partitioned molecule (19,20). Albertsson (19) found that the partition coefficient of various proteins and viruses depended on the surface area of the partitioned molecules. The greater the molecular weight of the molecule, the greater the surface area.

The partitioning of charged biomolecules and proteins between these two aqueous polymer phases can be further altered or enhanced by changes in the ionic composition of the phase system (21–23). Several inorganic salts have differing affinities for the phases in a dextran–polyethylene glycol (two water-soluble polymers)–water system (24). Because of the partitioning of the salts, it is possible to manipulate the partitioning of a charged biomolecule. The bonding of hydrophobic groups (25–27) or biospecific ligands (19,28) to one or more of the polymers used in a phase system has also been shown to influence the partitioning of biomolecules and proteins.

Integral membrane proteins have a hydrophobic domain that interacts with the hydrophobic core of the membrane lipids and a hydrophilic domain that is in contact with the surrounding aqueous medium (29). Isolation of the hydrophobic membrane proteins is necessary fot their study. The phase system formed using two or more polymers dissolved in water is not effective in the extraction of these hydrophobic proteins because the proteins do not solubilize in water. However, the hydrophobic proteins can be dissolved into an aqueous environment with the addition of surfactant. Upon the introduction of surfactant above the CMC, the membrane lipids that are associated with the hydrophobic protein disengage from the protein and solubilize in the micelles. The hydrophobic portion of the monomer then interacts with the protein surface, leaving the hydrophilic portion of the surfactant to interact with the aqueous environment making the protein water-soluble (29). However, since the surfaces of the proteins and other membrane components are no longer exposed to the environment, the individual characteristics and properties of the proteins are masked, causing their separation from each other to become difficult (30).

It is known that some detergents partition unequally among polymer two-phase systems (29). Therefore, separation of hydrophobic membrane components is made possible through the addition of surfactants to an aqueous two-phase system formed by the solubilization of polymer in water. The surfactant solubilizes the protein and then partitions itself among the phases formed by the presence of two or more dissimilar polymers. Albertsson (31) separated phospholipase A1, a hydrophobic membrane protein, from *Esherichia coli* by this procedure. In that work, a nonionic surfactant was added to a dextran—polyethylene glycol—water system. Nonionic surfactant is used because it is generally more mild than ionic surfactants and therefore more likely to preserve the biological activity of the protein.

The cloud point phenomenon of nonionic surfactants can be used to eliminate the need for the addition of polymers to isolate the hydrophobic proteins (32). To preserve the biological activity of the hydrophobic proteins, the nonionic surfactant to be used must have a cloud point below the temperature at which the material becomes biologically inactive. To correct the problem of too high a cloud point, it might be possible to add inorganic salts or other substances that lower the cloud point of the solution.

Only simple organic solutes were studied in this work. However, the technique should be applicable to biochemicals as well.

## III.  EXPERIMENTAL

### A.  Materials

Octylphenoxypoly(ethyleneoxy)ethanol with an average of 7.2 mol [OP(EO)$_{7.2}$:   trade name CA-620, GAF Corp.] and 5 mol [OP(EO)$_5$:   trade name CA-520, GAF Corp.] of ethylene oxide per mol of surfactant molecule were the two nonionic surfactants used in the study. These were used as received. The 4-tert-butylphenol (TBP) from Aldrich Chemical was 99% pure. Before use, the TBP was recrystallized from water and ethanol and dried at low heat. The hexanol and octanol, both obtained from Aldrich Chemical, were >98% pure and used as received. The water was distilled and deionized.

### B.  Methods

Several identical test tubes containing an aqueous solution with nonionic surfactant and a solute were placed in an isothermal water bath. Once phase separation had occurred, the fractional volume of each phase was determined by measuring the height of the phase. Samples of the upper phase were then withdrawn and, when possible, the lower phase (coacervate phase) was also sampled. The surfactant and solute concentrations in the samples were measured until equilibrium was reached, usually within 48 hr. The nonionic surfactant concentrations were analyzed with a Bausch and Lomb Spectronic 1001 UV spectrometer at 224 nm. The alcohols were analyzed on a Varian 3300 gas chromatograph with a flame ionization detector. The TBP was separated from the surfactant by reversed phase high-performance liquid chromatography before being analyzed by a UV detector.

## IV.  RESULTS

The initial surfactant-to-solute ratio was approximately 5:1.
Data on the fractional volume of the coacervate phase and the
compositions of the dilute and coacervate phases at equilibrium
are shown in Table 1 and Figs. 3–6.  By combining the phase
volume data with the concentrations of the components in each
of the phases, the amount of each component in a phase was
calculated.  The fraction of the total amount of each component
found in the coacervate phase is presented in Table 2.

Table 3 and Figs 3 and 5 present the ratio of the surfactant
and solute concentrations in the coacervate phase to that in the
dilute phase and to that in the initial solute or feed stream.  We
define these values as partition ratios.

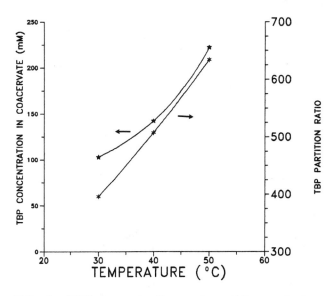

FIG. 3    TBP concentration and partition ratio in coacervate.

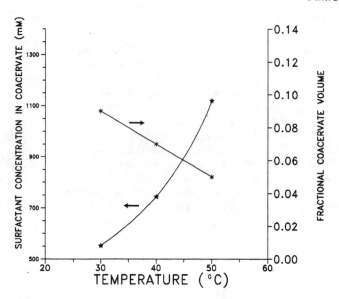

FIG. 4   Surfactant concentration in coacervate and fractional
volume of coacervate phase in system with TBP.

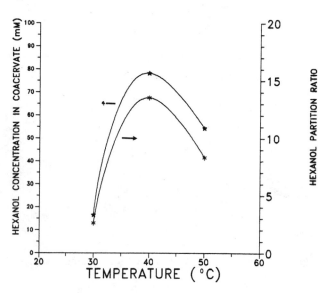

FIG. 5   Hexanol concentration and partition ratio in coacervate.

**TABLE 1** Data on Efficiency of Liquid-Coacervate Extraction

| System | Temperature (°C) | Fractional coacervate volume | [Surfactant] (mM) | | [Solute] (mM) | |
|---|---|---|---|---|---|---|
| | | | Dilute phase | Coacervate phase | Dilute phase | Coacervate phase |
| OP(EO)$_{7.2}$/hexanol | 30 | 0.20 | 4.84 | 206.0 | 6.42 | 16.5 |
| | 40 | 0.10 | 3.46 | 490.0 | 5.78 | 78.2 |
| | 50 | 0.08 | 4.83 | 635.0 | 6.54 | 54.3 |
| OP(EO)$_{7.2}$/octanol | 30 | 0.13 | 0.64 | 513.0 | 0.69 | 64.9 |
| OP(EO)$_5$/TBP | 30 | 0.09 | 0.32 | 553.0 | 0.26 | 103.0 |
| | 40 | 0.07 | 0.66 | 744.0 | 0.28 | 142.0 |
| | 50 | 0.05 | 0.30 | 1120.0 | 0.35 | 222.0 |

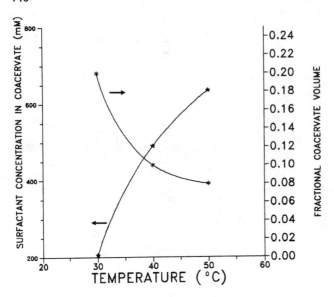

FIG. 6  Surfactant concentration in coacervate and fractional
volume of coacervate phase in system with hexanol.

TABLE 2    Fractional Removal of Components

| System | Temperature (°C) | Fraction of surfactant in coacervate | Fraction of solute in coacervate |
|---|---|---|---|
| OP(EO)$_{7.2}$/ hexanol | 30 | 0.91 | 0.39 |
| | 40 | 0.94 | 0.60 |
| | 50 | 0.92 | 0.42 |
| OP(EO)$_{7.2}$/ octanol | 30 | 0.99 | 0.93 |
| OP(EO)$_5$/ TBP | 30 | 0.99 | 0.97 |
| | 40 | 0.99 | 0.97 |
| | 50 | 0.99 | 0.97 |

TABLE 3 Partition Ratios of Components

| System | Temperature (°C) | $\dfrac{[\text{Surfactant}]_C}{[\text{Surfactant}]_D}$ | $\dfrac{[\text{Solute}]_C}{[\text{Solute}]_D}$ | $\dfrac{[\text{Solute}]_C}{[\text{Solute}]_F}$ |
|---|---|---|---|---|
| OP(EO)$_{7.2}$/hexanol | 30 | 42.6 | 2.6 | 1.95 |
| | 40 | 142.0 | 13.5 | 6.01 |
| | 50 | 131.0 | 8.3 | 5.24 |
| OP(EO)$_{7.2}$/octanol | 30 | 802.0 | 94.1 | 7.18 |
| OP(EO)$_5$/TBP | 30 | 1730.0 | 396.0 | 10.8 |
| | 40 | 1130.0 | 507.0 | 13.9 |
| | 50 | 3730.0 | 634.0 | 19.4 |

Subscripts: C, coacervate phase; D, dilute phase; F, feed.

## V.  DISCUSSION

As seen in Tables 2 and 3 and Fig. 3, for streams containing
TBP, 97% of the solute is removed in the coacervate, with the
solute concentration in the coacervate being 634 times as concentrated as in the dilute stream and 19.4 times as concentrated
as in the feed.  This shows that very high separation factors and
recoveries can be attained in one stage.  This is not an optimized
system; these were feasibility studies.  Surfactant structure
optimization and staging offer promise for a much better
performance.

As seen in Table 2, 91–99% of the surfactant was in the
coacervate phase.  This high partition ratio for the surfactant is
necessary for effective separations because additional surfactant
in the dilute phase is present as micelles.  The resulting solubilization in the dilute phase reduces the partition ratio and is
deleterious to the separation.

As seen in Tables 2 and 3, octanol partitions more effectively
into the coacervate phase than hexanol.  In addition, a higher
fraction of the original octanol ends up in the coacervate (93%
vs. 39%).  The exact structure of the surfactant in the coacervate
is not known.  However, the aggregate structure probably consists of the surfactant hydrocarbon chains intertwining, removing
themselves from aqueous environment, and the hydrophilic groups
covering the surface of this hydrophobic region.  Structures
such as micelles (spherical, rodlike, or lamellar) or vesicles are
feasible structures (the coacervate does not polarize light, so it
is not a liquid crystal).  For example, the nonideality of mixing
of ionic and nonionic surfactants in the coacervate is very similar to that associated with formation of mixed micelles for the
same system (2).  In this case, it is reasonable that the tendency
of an organic solute to solubilize in a micelle would be indicative
of the tendency to distribute itself into the coacervate phase.
As discussed in Chapter 1, a more hydrophobic member of a
homologous series has a greater tendency to solubilize in micelles.
Therefore, the higher partition ratio into the coacervate for
octanol compared to hexanol is reasonable.

The system $OP(EO)_{7.2}/TBP$ did not exhibit good phase separation characteristics, so a different surfactant was used for
the phenolic removal.  The water solubility of TBP and octanol
are similar.  At 30°C, the TBP exhibits higher partition ratios
(Table 3) and fraction removed (Table 2) than octanol.  Both
TBP and octanol exhibit more effective removal than the highly
water-soluble hexanol.

From Table 1 and Figs. 4 and 6, as the temperature of a system was increased, the volume fraction of the coacervate phase decreased and the surfactant concentration in the coacervate increased. This is because a temperature increase causes the system to be further from the critical point (cloud point), resulting in increasing dissimilarity between coacervate and dilute phases. Since an increase in the degree of ethoxylation of the surfactant causes an increase in cloud point (1), this proximity to the critical point can also explain the lower fractional phase volume for the coacervate for the $OP(EO)_5$ surfactant system compared to the $OP(EO)_{7.2}$ system.

As seen in Tables 1 and 3 and Figs. 3 and 5, as the temperature increases, the partition ratio and the solute fraction in the coacervate increase for TBP but show a maximum for hexanol. In some systems at least, there appears to be an optimum temperature for this separation. This implies that at a specified temperature there is an optimum surfactant structure (and associated cloud point) for this separation.

Liquid-coacervate extraction has been shown to be effective for two representative alcohols and a phenol, and has promise to be effective in the extraction and concentration of biological materials. This method would be ideally suited to this application since both phases are aqueous and therefore much less likely to denature or harm the biological activity of these substances. Also, separation and concentration can be achieved at low, energy-saving, less biologically hazardous temperatures.

## REFERENCES

1. M. J. Rosen, *Surfactants and Interfacial Phenomena*, Wiley, New York, 1978, p. 139.
2. O. E. Yoesting and J. F. Scamehorn, *Colloid Polym. Sci.*, *264*: 148 (1986).
3. M. Corti, C. Minero, and V. Degiorgio, *J. Phys. Chem.*, *88*: 309 (1984).
4. D. J. Mitchell, G. J. T. Tiddy, L. Waring, T. Bostock, and M. P. McDonald, *J. Chem. Soc., Faraday Trans. I*, *79*: 975 (1983).
5. J. Doren and J. Goldfarb, *J. Colloid Interf. Sci.*, *32*: 67 (1970).
6. M. Corti, V. Degiorgio, and M. Zulauf, *Phys. Rev. Lett.*, *48*: 1617 (1982).
7. L. S. C. Wan, *J. Am. Oil Chem. Soc.*, *60*: 1359 (1983).

8.  L. Sepulevda and F. MacRitchie, *J. Colloid Interf. Sci.*, *28*: 19 (1968).
9.  J. Favre and D. E. Pettijohn, *Eur. J. Biochem.*, *3*: 33 (1967).
10. G. Johansson, *J. Chromatogr.*, *150*: 63 (1978).
11. D. Kessel, *Biochem. Biophys. Acta*, *678*: 245 (1981).
12. D. E. Pettijohn, *Eur. J. Biochem.*, *3*: 25 (1967).
13. L. Rudin, *Biochem. Biophys. Acta*, *134*: 199 (1967).
14. L. Walter and .F. W. Selby, *Biochem. Biophys. Acta*, *148*: 517 (1967).
15. G. Johansson, *Biochem. Biophys. Acta*, *451*: 517 (1976).
16. A. Polson, G. M. Potgieter, J. F. Largier, G. E. F. Mears, and F. J. Joubert, *Biochem. Biophys. Acta*, *82*: 463 (1964).
17. G. Johansson, A. Hartman, and P. A. Albertsson, *Eur. J. Biochem.*, *33*: 379 (1973).
18. M. Zeppezauer and S. Brishammar, *Biochem. Biophys. Acta*, *94*: 581 (1965).
19. P. A. Albertsson, *Partition of Cell Particles and Macromolecules*, 3rd ed., Wiley, New York, 1986, Chap. 4.
20. P. A. Albertsson, *Meth. Biochem. Anal.*, *10*: 229 (1962).
21. P. A. Albertsson, *Biochem. Biophys. Acta*, *103*: 1 (1965).
22. C. M. Ballard, M. H. W. Roberts, and J. P. Dickinson, *Biochem. Biophys. Acta*, *582*: 89 (1979).
23. G. Johansson, *Biochem. Biophys. Acta*, *222*: 381 (1970).
24. G. Johansson, *Biochem. Biophys. Acta*, *221*: 387 (1970).
25. V. P. Shanbhag and G. Johansson, *Biochem. Biophys. Res. Commun.*, *61*: 1141 (1974).
26. V. P. Shanbhag and C. G. Axelsson, *Eur. J. Biochem.*, *60*: 17 (1975).
27. H. Westrin, P. A. Albertsson, and G. Johnsson, *Biochem. Biophys. Acta*, *436*: 696 (1976).
28. G. Johansson, G. Kopperschlager, and P. A. Albertsson, *Eur. J. Biochem.*, *131*: 589 (1983).
29. P. A. Albertsson and B. J. Andersson, *J. Chromatogr.*, *215*: 131 (1981).
30. P. Svensson, W. Schroder, H. E. Akerlund, and P. A. Albertsson, *J. Chromatogr.*, *323*: 363 (1985).
31. P. A. Albertsson, *Biochemistry*, *12*: 2525 (1973).
32. R. Heusch, *In. Z. Biotech.*, *3*: 2 (1986).

# III
## SEPARATIONS BASED ON ADSORPTION

# 7

# Adsorbed Surfactant Bilayers as Two-Dimensional Solvents: Admicellar-Enhanced Chromatography

JEFFREY H. HARWELL* Institute for Applied Surfactant Research, University of Oklahoma, Norman, Oklahoma

EDGAR A. O'REAR School of Chemical Engineering and Materials Science, University of Oklahoma, Norman, Oklahoma

*Present affiliation: Directorate for Engineering, Division of Chemistry, Biochemistry, and Thermal Engineering, National Science Foundation, Washington, D.C.

This material is based on work supported by the National Science Foundation under Grant No. CPE-8318864.

SYNOPSIS

Admicellar-enhanced chromatography (AEC) is a new fixed-bed
separation process based on the use of surfactants to induce
the adsorption of a solute from an aqueous stream. In AEC,
an adsorbed layer of surfactant is formed reversibly on the
surface of a packing material. Solute molecules from a process
stream partition into the surfactant layer in a phenomenon
called adsolubilization, which is a surface analog of solubilization. When the surfactant layer—or, equivalently, the bed—
becomes saturated with the solute, the surfactant layer with
the solute is stripped from the packing by a change of pH in
the inlet stream to the bed. This can be readily accomplished
because of the nature of the surface-charging phenomenon that
is used to induce the formation of the surfactant layer in the
first place. The concentration of the solute in the stream of
material stripped from the bed can be adjusted by controlling
the pH of the stripping stream, and may be much higher than
the original concentration of the solute in the process stream.
AEC has now been used in bench scale applications to concentrate a phenolic compound from an aqueous stream, to separate
isomers of an alcohol, and shows promise as an industrial scale
method for separating optical isomers.


I.  INTRODUCTION

The advent of gene splicing, monoclonal antibody technology,
production of biofuels for renewable energy sources, and broad
environmental awareness has brought new challenges in the
isolation of dilute aqueous product molecules or pollutants (1).
Surfactant-based or surfactant-augmented methods show great
promise in meeting these needs. Conceivable advantages of
surfactant techniques include flexibility of design—given the
large variety of known surfactant molecules, the possibility of
molecular design of a chemical surfactant to effect a particular
separation, and greater versatility of capital equipment as well as
relative ease of future process modification by simply switching
the surfactant rather than the process equipment. Additionally,
surfactants function at relatively low concentration, are easy to
work with, have well-characterized fundamental properties, and
many are nontoxic and biodegradable.

    All molecular separations require multiple "phases" or
spatial regions with interfaces or boundaries across which

disparate fluxes must occur. In surfactant-aided separations, surfactant molecules generally cause the creation of these new phases or regions which favor the incorporation of particular species or types of species. At the same time, however, surfactant-based methods have often brought to the forefront another challenge—the requirement to isolate the separate "phases," since the micellar pseudophase may actually be molecularly dispersed in the external phase. For example, the observance of selective solubilization by Nagarajan and Ruckenstein (2) lead conceptually to the use of surfactant micelles for multicomponent separation. Practical implementation, however, has generally remained for the development of techniques capable of addressing the issue of phase separation. One such technique, described in Chapters 1 and 2 of this book, is ultra-filtration of a micellar solution, which was first proposed by Leung at Union Carbide (3) but has subsequently matured as micellar enhanced ultrafiltration (MEUF) under the direction of Scamehorn, Christian, and coworkers (4,5) at the University of Oklahoma. In this method, ultrafiltration effects separation of the dispersed phase by concentrating the micelles with the solubilizates.

Surfactant-based or surfactant aggregate-based separations may be divided into two arbitrary categories. First is a special case of either the concentrating of a single species or the non-selective removal of similar species from dilute solution (type I) while the second is the selective removal of a target species from a multicomponent mixture (type II). Thus, the removal of all transition metals from a wastewater stream belongs to the type I category while the removal of silver ions from a process solution of transition metal ions is an example of type II separations. The focus to date for MEUF has been primarily on type I separations, though experiments currently underway begin to explore the potential of MEUF for type II separations.

Admicellar-enhanced chromatography (AEC) utilizes adsorbed monolayer or bilayer surfactant aggregates called, respectively, hemimicelles or admicelles and the phenomenon of adsolubilization. In adsolubilization, species partition between the adsorbed surfactant phase and the bulk or supernatant solution. Figure 1 shows AEC being used for a type I separation. The issue of phase separation is addressed in admicellar enhanced chromatography by the inherent immobilization of the surfactant aggregates on the substrate surface. The solid phase facilitates the removal of the supernatant and its solutes while it retains the admicellar surfactant phase and any associated species.

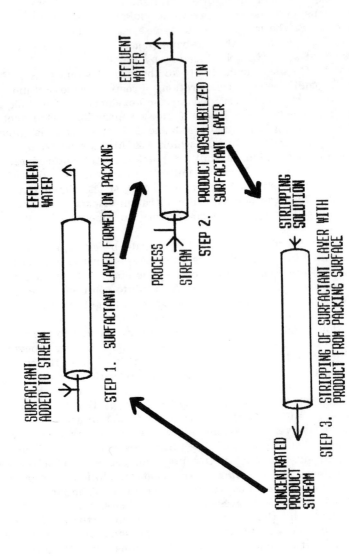

FIG. 1   Concentration of a compound from dilute solution by admicellar-enhanced chromatography.

This chapter describes the underlying phenomena of admicellar-enhanced chromatography, the basic steps of the process, and its application to separations of type I or type II. In the succeeding chapter, the phenomenon of adsolubilization is examined with emphasis on the preparation of thin films to construct unique chromatographic or ion exchange packings.


## II.  ADMICELLE FORMATION

The two principal physical phenomena that underlie AEC are the reversible formation of surfactant aggregates at a solid-liquid interface and the concentration of solutes at the interface through their incorporation into the admicelle structure. That surfactants tend to aggregate at the solid-liquid interface is known. Factors controlling the adsorption include type of surfactant, solvent, and substrate as well as pH, concentration of added electrolyte, binding of gegenions, and the presence of adsolubilizates. A typical adsorption isotherm for an ionic surfactant on a metal oxide is shown in Fig. 2. At low coverages in the Henry's law region, individual surfactant monomers adsorb onto the substrate. The region of the isotherm in which this is the predominant mode of surfactant adsorption is frequently referred to as region I (6). The amount adsorbed in region I increases linearly with concentration of the surfactant solution in equilibrium with the surface. At somewhat higher coverages, cooperative effects and association of surfactant on the surface begin; this is observed as an increase in the slope of the adsorption isotherm. Adsorbed monolayer surfactant aggregates have been called hemimicelles, while bilayer structures are distinguished by the name of admicelles. The change in slope reflects greater surfactant adsorption for a small increment of surfactant concentration in the supernatant and signals that the critical admicelle concentration (CAC) (7) or hemimicelle concentration (HMC) (8) has been reached. This region of greatly increased slope is generally referred to as region II. Here, formation of aggregates on patches of the surface is the principal mechanism of surfactant adsorption. Monomers are still adsorbing in an unaggregated state on less energetic patches of the surface, but their contribution to the total adsorption is negligible. A region of decreased slope, called region III, is distinguished by some workers. Whether or not there is a change in adsorption mechanism at the region II/region III transition is a matter of some controversy

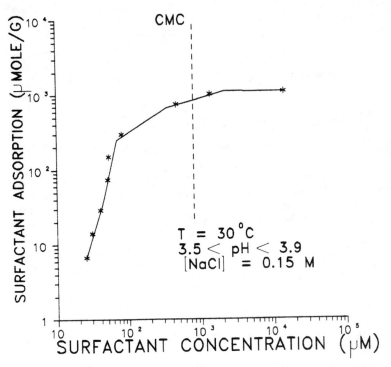

FIG. 2    Typical surfactant adsorption isotherm.    (Reprinted by permission from Ref. 9.)

(9).  Whether the aggregates forming in this part of the iso-therm are admicelles, hemimicelles, or a combination of the two also has been the subject of opposing theories (9).   Among these, it has been suggested that only monolayer structures form (10), that bilayer formation begins only after completion of the monolayer (11), or that dispersed monolayer and bilayer structures nucleate and grow simultaneously in regions II and III (6,9).  Regardless of their exact nature, sufficient aggre-gates of surfactant can be created to demonstrate the features of admicellar chromatography.

From Fig. 2, it is seen that adsorption rises and plateaus as the surfactant concentration is increased.  While plateau adsorption may occur from attaining the CMC of surfactant in the supernatant (6), it may also result from surface saturation

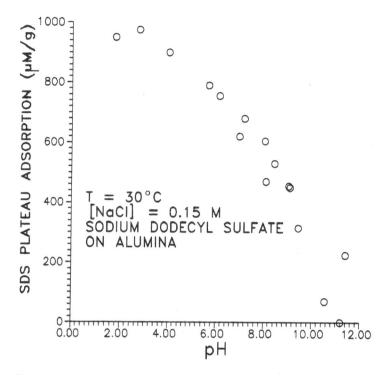

FIG. 3    Plateau adsorption as a function of pH.    (Reprinted by permission from Ref. 9.)

in systems applicable to admicellar chromatography (9).   Reflecting strong dependence on surface charge, plateau adsorption is highly dependent on pH of the system, as illustrated in Fig. 3. For an anionic surfactant in aqueous solution, adsorption increases as pH is lowered.   The greater acidity causes protonation of the surface (12) and enhanced Coulombic attraction of the surfactant at the interface.   Conversely, raising the pH leads to desorption of anionic surfactant.   Key steps in admicellar enhanced chromatography rely on the reversible formation of the aggregates, which is controlled primarily by such shifts in acidity or alkalinity to affect a crossing of the point of zero charge (pzc) and thus a desorption of the admicelles.

Finally, it should be noted that near the pzc, the plateau becomes less distinct.   In fact, as illustrated in Fig. 4 (13), even well below the pzc the plateau effectively ceases to exist.

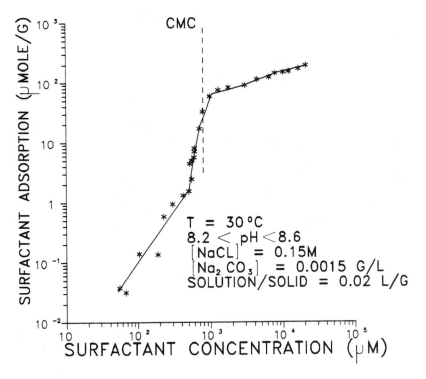

FIG. 4   Surfactant adsorption isotherm near the pzc.   (Reprinted by permission from Ref. 13.)

A probable explanation for this phenomenon is that such a small fraction of the surface is covered when the surfactant concentration reaches the CMC, that even the small changes in chemical potential near the CMC, relative to changes in the concentration and morphology of the micelles, are sufficient to induce formation of additional aggregates at the solid-solution interface.

## III.  ADSOLUBILIZATION

Nonsurfactant molecules and ions can be concentrated or associated with micelles in solution in a process called solubilization. A similar but less well-known phenomenon occurs with admicelles

and hemimicelles in adsolubilization. Adsolubilization may be
described as the localization of non-surface-active molecules at
a liquid-solid interface through the assistance of adsorbed surfactant molecules. Partitioning of species into admicelles
appears to be very general. To date, adsolubilizates in our
labs have included low molecular weight linear alkanes (14),
aliphatic alcohols (15,16), and phenols (17). Generally, the
amount of adsolubilization rises with increasing surfactant adsorption (i.e., the extent of the admicellar phase) and with
increasing concentration of the adsolubilizate in the supernatant. However, saturation of the surfactant aggregates also
occurs. The limiting ratio of surfactant to adsolubilizate for
the saturated and aromatic hydrocarbons tested approaches
values of 2:1. Much higher ratios are seen for alcohols at
low coverage of surfactant (14). As surfactant coverage nears
completion the moles of alcohol adsolubilized to moles of surfactant approaches values seen for micelles. Higher ratios of
alcohol to surfactant which have been observed at low coverage
could be explained by patchwise adsorption with molecules of
alcohol surrounding and within the separate surfactant aggregates.

## IV.   BASIC STEPS IN ADMICELLAR-ENHANCED
##       CHROMATOGRAPHY

Admicellar-enhanced chromatography can be operated in a
fashion analogous to conventional liquid chromatography, but it
exhibits its greatest potential when operated in a unique multi-
step, cyclical fashion (16−18). Conceptually, a cycle consists
of a bed preparation step, adsolubilization of target species,
and finally an admicelle-stripping step to release surfactant
and product from the bed (Fig. 1). In the bed preparation
phase, surfactant is introduced into the packed column under
conditions that favor the formation of admicelles on the packing
(e.g., an acidic pH for an anionic surfactant). If necessary,
the bed can be prewashed with dilute acid or base solution to
obtain the desired pH. The packing material is usually a
porous, inorganic solid (metal oxide) support which provides
large interfacial area for the admicellar phase and consequently
higher bed capacity.
     As the relatively concentrated surfactant feed enters the
column, surfactant is removed from the solution and deposited
onto the bed to form the admicelles. After a while the column
becomes saturated and the concentration of surfactant in the

exiting stream rises rapidly toward the feed concentration. The
product stream to be separated or concentrated is now admitted
to the column; some surfactant must continue to be added in
the feed in order to maintain the surface aggregate structures
and not shift the admicellar equilibrium toward desorption. In
practice, these first two steps may be carried out simultaneously
to reduce surfactant usage. With proper selection of surfactant
and operating conditions, adsolubilization or even selective
adsolubilization can be achieved. The effluent stream is moni-
tored and the feed continued until the bed is saturated with
respect to the adsolubilizate. In the third and final phase, the
bed-stripping phase, an acidic or alkaline pulse is fed into the
column to force the desorption of the surfactant with its adsolubilizate. Results from a type I separation with a phenolic
have shown an increase in concentration by two- to threefold
in a single stage with low energy requirements. Though the
process is still not optimized, cleaner elution of the product
from the bed has been obtained by beginning the bed-stripping
step with the adsolubilizate wave only partially through the
column (Fig. 5).

Special note should be taken of the mechanism by which
the concentration effect is achieved. The injection of the stripping stream (just an aqueous solution with a pH above the pzc
of the packing) sets up several chromatographic waves of varying velocities. In any multicomponent chromatographic system,
only one component can disappear across a given wave (19).
Since the stripping stream contains neither surfactant nor
product, the last wave to elute from the column must be the
wave of the change in the pH; all the product and all the surfactant will elute ahead of this pH change wave. The concentration of the product in the eluent will depend on the velocity
of the pH wave. If the pH wave is infinitely slow, the product
will be desorbed into a volume of water larger than it was adsorbed from, and no concentration effect will be observed. If
the velocity of the pH change wave is equal to the velocity of
the wave of product as it propagated through the bed during
the adsolubilization step, the final concentration will be equal
to the initial concentration. If, however, the velocity of the
pH change wave is greater than the velocity of the product
wave during the adsolubilization step, then the product will be
desorbed into a smaller volume of water than it was adsorbed
from, and a concentration increase will be observed. The
velocity of the pH change wave will increase as the pH of the
stripping solution is increased further and further from the

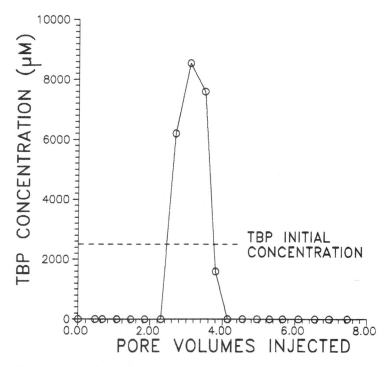

FIG. 5    Recovery of 4-*tert*-butylphenol from aqueous solution by AEC.

pzc of the packing.  As the pH of the stripping solution moves very far from the pzc, the velocity of the pH change wave begins to approach the velocity of the bulk fluid solution; as this happens, the volume of solution into which the product is desorbed begins to approach zero, and the concentration of the product in the eluent increases dramatically.  This phenomenon can be thought of as analogous to frontal displacement chromatography.  It is the mechanism by which the product can be concentrated.  Of course, the notion of using a highly basic or highly acidic stream for stripping the bed might seem to rule out the application of this technology to bioseparations.  In fact, since the product and the surfactant elute before the pH wave, the product is never exposed to the pH of the stripping solution.

Another unusual feature of admicellar-enhanced chromatography is that following the completion of one cycle, the bed is available for a repeat of the previous cycle or for a completely different separation. Indeed, because the surface charge crosses the point of zero charge over the course of a cycle, one effective method of operation might alternate anionic and cationic surfactants, whether for the same or different separations.

Selection of the packing and the type of surfactant for AEC is usually made in tandem. Matching the ionic surfactant with the adsorbent is based on the point of zero charge for the adsorbent relative to neutral pH. For a packing material with a pzc above pH 7, an anionic surfactant would be chosen, while for an adsorbent with a pzc below pH 7, a cationic surfactant would be the surfactant of choice. The rationale for these groupings is that addition of the potential determining ions, $H^+$ and $OH^-$, to enhance the adsorption of surfactant will also tend to move the system toward neutral pH and away from a harsh acidic or alkaline environment. Moreover, operating near neutral pH reduces the amount of acid or base needed for the process. As a result, it is possible to operate an AEC column under very mild condition, even while using pH changes to form and strip the admicelles.

There are a great many different known surfactants, so that it may be possible with time to develop rules for the design of surfactant for specific separation problems in a manner analogous to affinity chromatography. Equilibrium or batch adsolubilization experiments can be conducted to study prospective surfactants. A batch single stage separation is demonstrated in Fig. 6 to illustrate the evaluation of a surfactant in this manner and the concept of selective adsolubilization.

As illustrated in Fig. 6, a binary mixture of compounds A and B in solution is added to the preformed admicelles. Compound A partitions preferentially into the surfactant bilayer and is thus retained as the B-rich supernatant is removed. Reversing the surfactant charge on the adsorbate yields a solution rich in A. Multiple stages should enrich the separation so that in practice the chromatographic mode is more desirable. It is expected that the surfactant structure will be a primary factor in affecting selectivity of adsolubilizates. Other factors that can be varied are surfactant charge and coverage of the bed or packing density of the surfactant. It has been shown in equilibrium experiments (16) that the selectivity between aliphatic alcohols A and B increases as the surface structure becomes more tightly packed. This suggests that steric

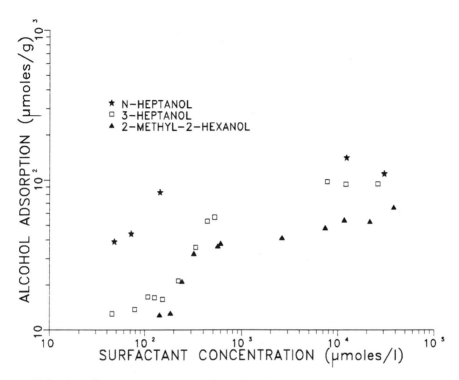

FIG. 6    Single-stage separation of A from B in dilute aqueous
solution using admicelles.

interactions between surfactant and adsolubilizate determine
selectivity.  This observation is consistent with the recent
measurement of very high microviscosites in the aggregates
obtained from fluorescence probes (20).  A commonly used
method of purification of organic and inorganic compounds is
crystallization.  Crystallization especially imparts selectivity
due to the close molecular interaction and high degree of order
within the crystal.  In a similar fashion, a relatively high
degree of order in the organized bilayer assembly may help
AEC to show related selectivity.
      A hypothetical separation can be used to illustrate the
selection of a surfactant for a particular separation problem.
Consider the separation of mirror image isomers or enantiomers.
These isomers have identical physical properties, so that their

**TABLE 1**   Adsolubilization of (±) Camphor in Adsorbed
Sodium Deoxycholate on Aluminum Oxide

| Feed concentration of deoxycholate (mM) | | Adsolubilization (μmol/g of alumina) | |
|---|---|---|---|
| (+) | (−) | (+) | (−) |
| 10 | 10 | 9.0 | 9.5 |
| 15 | 15 | 19.5 | 18.5 |
| 20 | 20 | 20.5 | 21.5 |
| 30 | 30 | 30.0 | 33.5 |
| 40 | 40 | 29.5 | 31.5 |
| 50 | 50 | 27.8 | 30.8 |

separation is very difficult and is often effected by reaction
with another pure optically active compound to form a diastereo-
mer.   If compounds A and B are the +/− forms in a mixture,
then the surfactant to be selected must contain a chiral center
and be a pure optical isomer.   As a result, one of the enantiomers in the racemate can be expected in principle to adsolubilize
to a greater degree due to a more favorable "diastereomeric
association" without the formation of a covalent bond.   Classical
resolution of optical isomers by crystallization results from the
requirements of the highly ordered crystal structure to impart
selectivity.   The high degree of order believed to exist in the
admicelle may permit AEC to be applied effectively in the resolution of optical isomers.   Small but discernible differences in the
absolubilization of camphor in equilibrium experiments were
found (Table 1) using sodium deoxycholate on aluminum oxide.
     At present, however, surfactant selections must largely be
based on empirical tests.   For instance, the single component
adsolubilization of three aliphatic alcohols in an aqueous system
of SDS admicelles gave adsolubilization results shown in Fig. 7.
Provided the system is ideal or nearly so, then we should be
able to separate alcohol A or B from alcohol C by operating in a
high surfactant concentration regime.   Alternatively, alcohol A
could be separated from either alcohol B or alcohol C if the
system is operated at low surfactant conditions.   It is easy to
imagine two-stage arrangements that would function to split a
ternary mixture into the individual components.

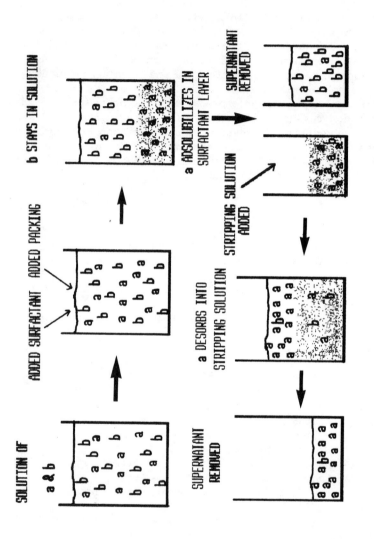

FIG. 7 Effect of branching on the adsolubilization of aliphatic alcohols. (From Ref. 16.)

Aside from surfactant structure, functional groups and molecular structures of the adsolubilizates will be significant design factors in AEC. For aliphatic alcohols, adsolubilization increases with chain length and decreases with branching, as also illustrated in Fig. 7 (16). These trends suggest a relationship to hydrophobicity of the adsolubilizate in these aqueous systems or possibly to the steric constraints within the admicelle. Important areas for future work on admicellar-enhanced chromatography are experimental data and partition relationships based on molecular structures of surfactant and adsolubilizate, comparison and possible interprediction of solubilization and adsolubilization results, exploration of the use of mixed micelles or other solvents and cosolvents, and optimization of operational methods for AEC. With simple or complex modes of operation possible, admicellar-enhanced chromatography is a powerful surfactant-based separation technology.

## REFERENCES

1. National Research Council, *Separation and Purification: Critical Needs and Opportunities*, National Academy Press, Washington, D.C., 1987.
2. R. Nagarjan and E. Ruckenstein, *Sep. Sci. Technol.*, *16*: 1429 (1981).
3. P. S. Leung, in *Ultrafiltration Membranes and Applications* (A. R. Cooper, ed.), Plenum Press, New York, 1979, p. 415.
4. R. O. Dunn, Jr., J. F. Scamehorn, and S. D. Christian, *Sep. Sci. Technol.*, *20*: 257 (1985).
5. J. F. Scamehorn, R. T. Ellington, S. D. Christian, B. W. Penny, R. O. Dunn, and S. N. Bhat, *AICHE Symp. Ser.*, *250*: 45 (1986).
6. J. F. Scamehorn, R. S. Schechter, and W. H. Wade, *J. Colloid Interf. Sci.*, *85*: 463 (1982).
7. J. H. Harwell, J. C. Hoskins, R. S. Schechter, and W. H. Wade, *Langmuir*, *1*: 251 (1985).
8. P. Somasundaran and D. W. Fuerstenau, *J. Phys. Chem.*, *70*: 90 (1986).
9. D. Bitting and J. H. Harwell, *Langmuir*, *3*: 500 (1987).
10. P. Sircuse and P. Somasundaran, *Colloid Surf.*, *26*: 55 (1987).
11. P. D. Bisio, J. G. Cartledge, W. H. Kessom, and C. J. Radke, *J. Colloid Interf. Sci.*, *78*: 225 (1980).

12. J. A. Davis, R. O. James, and J. O. Leckie, *J. Colloid Interf. Sci.*, *64*: 480 (1978).
13. B. L. Roberts, J. H. Harwell, and J. F. Scamehorn, *ACS Symp. Ser.*, Vol. 311, 1986, p. 200.
14. M. A. Yeskie and J. H. Harwell (in preparation).
15. M. A. Yeskie and J. H. Harwell (in preparation).
16. J. W. Barton, T. P. Fitzgerald, C. Lee, E. A. O'Rear, and J. H. Harwell, *Sep. Sci. Technol.*, *23*: 637 (1988).
17. C. R. Grout and J. H. Harwell, *Proc. of the Int. Congr. on Recent Advances in the Management of Hazardous and Toxic Wastes in the Process Industries* (in press).
18. T. P. Fitzgerald and J. H. Harwell, *AIChE Symp. Ser.*, *82*: 142 (1986).
19. F. G. Helfferich and G. Kline, *Multicomponent Chromatography: Theory of Interference*, Marcel Dekker, New York, 1970.
20. P. Chandar, P. Somasundaran, and N. Turro, *J. Phys. Chem.*, *117*: 31 (1987).

# 8

# Adsorbed Surfactant Bilayers as Two-Dimensional Solvents: Surface Modification by Thin-Film Formation

JENGYUE WU,* CHONLIN LEE, and JEFFREY H. HARWELL†
Institute for Applied Surfactant Research, University of
Oklahoma, Norman, Oklahoma

EDGAR A. O'REAR  School of Chemical Engineering and Materials
Science, University of Oklahoma, Norman, Oklahoma

*Present affiliations:*
*Division of Polymer Science, Institute for Research in Chemical Industry, Hsinchu, Taiwan.
†Directorate for Engineering, Division of Chemistry, Biochemistry, and Thermal Engineering, National Science Foundation, Washington, D.C.

SYNOPSIS

A new surface modification process is described in which polymers are formed at the surface of packing materials. The film
formation technique is conceptually a surface analog to emulsion
polymerization. By this method, a prototype, hybrid ion ex-
change-reversed phase chromatographic packing material has
been prepared and tested.

## I. INTRODUCTION

Today the most frequently used analytical separation technique
is partition chromatography, mainly because of its speed and
resolving power. However, because the entire field of separation technology continues to expand its capabilities for different
separations, packings with tailored properties are expected to
be required to meet special demands. This leads logically to
developments in the area of the surface modification chemistry
of packing in order to enhance selective adsorption of particular compounds. A common and established approach to enhanc-
ing selectivity is to control the packing formation process,
which affects the physical properties of the bulk material as
well as the surface properties. For example, the characteristics of alumina depend on the degree of dehydroxylation, which
can be optimized through thermal treatment (1). A more complex approach is the addition of chemical modifiers to increase
the concentration of the existing desired functional groups.
Because the variety of chemical functional groups that can be
adsorbed on two popularly used packings, silica and alumina,
are so diverse, a third type of surface modification becomes
attractive. This involves chemical treatment of the surface in
order to form new functionalities, such as a reversed phase
chromatographic packing, which can enhance the adsorption of
certain molecules.

A distinctive approach introduced recently (2) is to form a
thin film, using surfactants, on the surface of an inorganic

support.  This surface modification technique cannot only
alter the characteristics of the original support, but can
greatly improve the feasibility of attaching numerous functional groups on packing surfaces, such that many varieties
of application can be envisioned.  This chapter reviews the
film-forming process, concentrating on its application in surface modification for adsorbents.


## II.  FORMATION OF ULTRATHIN FILMS

The film-forming process is based on the formation of micelle-
like surfactant aggregates, which will be called admicelles (3,4),
at a solid-solution interface.  Just as in the phenomenon of
solubilization for micelles, the analogous behavior of partitioning
of organic solutes into admicelles has been termed adsolubilization.  These two phenomena, admicelle formation and adsolubilization, discussed in more detail in Chapter 7 of this book,
form the basis of a three-step process to construct a polymeric
thin film on a solid substrate via a low-energy process, as
illustrated in Fig. 1.  First, an ionic surfactant is adsorbed on
an oppositely charged surface by controlling adsorption parameters such as pH and salt concentration.  Second, monomers
are adsolubilized into the organic-like admicellar interior.  With
the addition of initiator and the supply of heat or light, the
polymerization is accomplished; this constitutes the final step
of the process for the in situ formation of ultrathin films.  The
final product of the process will not be a monomolecular layer
like the Langmuir-Blodgett film (5) but a bilayer.  With the
various combinations of surfactant and monomers possible in
the three-step construction of the product, a great variety of
feasible ways exist to form a film on a substrate and to construct films with desirable characteristics.  Additionally, films
can be formed on highly irregular surfaces, including the
extreme of forming the polymerized bilayer on the internal
surfaces of porous solids (6).  By choosing different combinations of surfactant and monomers, based on the functionalities
and end groups desired, an assortment of films can, in principle, be made for consideration in a particular application.

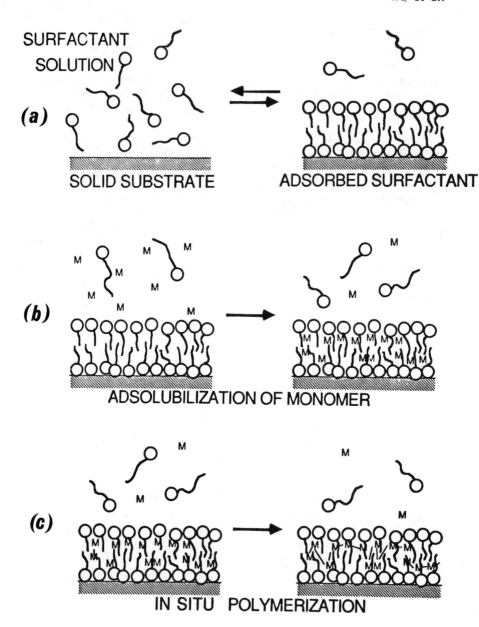

FIG. 1 A three-step process for constructing ultrathin polymeric films from physically adsorbed surfactant bilayers.
(Reprinted with permission from Ref. 6.)

## III. CHARACTERISTICS OF THE THIN FILM

Via the novel procedure described above, the formation of ultrathin polymer films on solid surfaces has been demonstrated (6)
by using such common materials as alumina powders, sodium
dodecyl sulfate, styrene, and the free radical initiator sodium
persulfate. Several techniques (7,8) have been applied to
characterize films formed by this system. Spectroscopic evidence for polymerization on the alumina powder surface was obtained by isolating, washing, and then drying the alumina
powder in vacuum, followed by extraction with tetrahydrofuran
(THF). In Fig. 2, the UV spectrum of the extract in THF is
compared with the spectrum of a known sample of polystyrene
(MW 2350, Waters Associates, Milford, Mass.) dissolved in THF.
Spectra of higher molecular weight polystyrenes were similar,
although a slight dependence on molecular weight was found for
the chromophore concentration-based extinction coefficient.

In the same manner, the conversion of styrene to polystyrene on alumina surfaces was measured with respect to reaction time (7). An initiator-concentration-dependent "induction
time," which is typical for free radical polymerization systems,
was observed to be about 10 min. The conversion reaches almost 50% after half an hour. However, with lower initiator concentration, the reaction time to reach 50% conversion could be
as long as 45 min. Similarly, by increasing initiator concentration, 90% conversion could be achieved within 35 min. Results
from these kinetic studies also indicate that the principal polymerization loci appear to be within the admicelles.

After quenching the reaction, followed by separation, rinsing, and drying of the alumina, it was possible to observe the
polystyrene and SDS on the surface of the alumina with FTIR.
Figure 3 shows transmission FTIR in the range of 2300−1700
$cm^{-1}$ for SDS on alumina in a system after a 5-min reaction and
after a 20-min reaction. Three characteristic polystyrene peaks
emerge at 1455, 1490, and 1600 $cm^{-1}$. The latter of these
appears as a shoulder on the 1625 $cm^{-1}$ SDS peak.

Nitrogen sorption, a conventional tool for measuring surface
areas and pore size distribution, has been employed (6) for
determining the thickness of the thin film formed on the interior
of a porous substrate, through a comparison of the pore size
distribution curves before and after film formation. Film thicknesses, estimated from tie lines (Fig. 4), are seen to vary from
1.8 to 0.4 nm. For a 55% bilayer surface coverage, as determined by observed surfactant adsorption densities, the overall

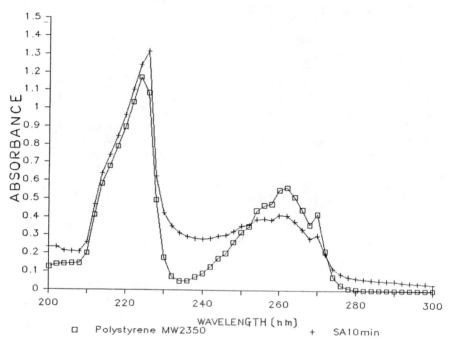

FIG. 2   Comparison of the UV spectrum of THF extract from the filmed alumina powder to a polystyrene sample of known molecular weight.   (Reprinted with permission from Ref. 2.)

average film thickness is 1.7 nm.   This is in good agreement with the thickness determined from the sorption data.   For larger pores, the apparent film thickness is greater.   This may reflect a more complete surface coverage of larger pores rather than a change in the morphology of the polymerized aggregate. More interestingly, while the BET surface area of the alumina was reduced by about 40% after the filming process, the pore size distributions obtained for both bare and filmed alumina have the same basic shape; this indicates that the thin film has reduced the pore diameters of all of the pores rather than simply plugging some of the pores.   This is a very important feature of this surface modification process to be applied to the making of unique packing materials for chromatographic separations.

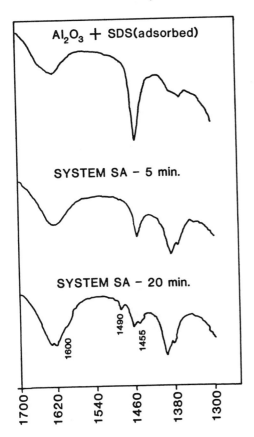

FIG. 3   Emergence of characteristic IR peaks for polystyrene
with increasing reaction time during the film-forming process.
(Reprinted with permission from Ref. 2.)

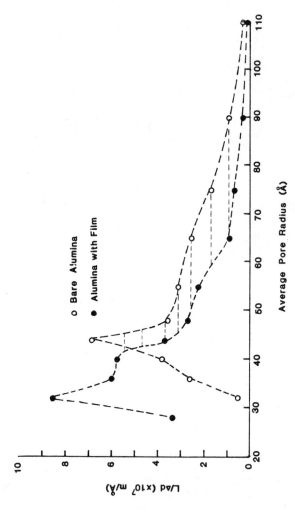

FIG. 4  Pore size distributions for the alumina powder before and after application of the thin film. Tielines show the change in pore diameters caused by application of the film. (Reprinted with permission from Ref. 6.)

## IV.  PREPARATION AND CHARACTERIZATION OF A MIXED-MODE PACKING MEDIUM

### A.  Overview of the Final Product

As mentioned above, polymerized films can be formed inside any pores of the solid into which surfactants will diffuse, so that high surface area packings are readily formed.  To achieve high selectivity, surface chemistry can be controlled through attaching outside coatings on the thin film or by choosing surfactant or monomers with either the desired functional groups or with a structure that imitates the solute to be resolved. For example, coated with the hydrophobic polymer film, high surface area alumina or silica powder might be used as packing material for reversed phase chromatography.  With the flexibility of designing surface functionalities, this filming process can potentially form a stationary phase carrying two or more types of functional groups with a controlled distribution.  A chromatograph operated in a mixed-mode fashion, such as simultaneous reversed phase/ion exchange chromatography, can be envisioned.

In addition, because the admicelles are self-assembling, we expect a high degree of molecular organization within the layer. These characteristics promise the ability to achieve remarkably high degrees of selectivity in a separation.  The incorporation of optically active surfactants in the polymerization steps might impart chirality to the resulting polymer film.  In this way, inexpensive and simple chiral packing might be used to resolve stereomers through differing interactions between asymmetric centers.  Apart from selectivity, improvement of the stability of the thin film on packings will play a decisive role in assuring the success of these applications of the films in separations.

This potential application of alumina powders coated with the ultrathin film based on the sodium dodecyl sulfate (SDS) and polystyrene complex is further explored here by testing the stability of the films and the plausibility of their use as a combined reversed phase/ion exchange medium.  In inspecting the structure of the SDS-polystyrene complex, as discussed before, two important aspects can be emphasized here.  One is the hydrophobic nature of the polymerized thin film surface that is obtained after washing the filmed alumina powders (6). The other is the retainment of cations, such as sodium ions, by the unwashed film.  These sodium ions are bound to the headgroups of the surfactant in this molecular complex by the electrostatic interactions. While cations are associated at least with the upper or second layer of surfactant molecules, exchangeable ions may also be associated with some of the surfactant ions in the lower or first layer of surfactant, since the charge on the alumina surface alone is insufficient to balance even the charge on the first layer surfactant in the admicelle (3). Not only can the now hydrophobic surface of the modified alumina accommodate the separation of organic-like compounds, but it can do so while retaining its ion exchange capacity.

After the adsolubilization step, but without the polymerization of the ultrathin film inside the adsorbed two-dimensional solvent on the alumina, the monomeric SDS molecule will readily desorb into an aqueous solution. SDS is fairly water-soluble above its Krafft temperature. Unless the film is polymerized, the introduction of an aqueous solution containing the ions to be exchanged will result in the elution of SDS from the solid substrate, as described in Chapter 7 of this book. Under such circumstances, the repeated ion exchange and regeneration steps will result in the gradual washing out of the SDS and a deterioration of the ability of an inorganic packing to retain hydrophobic solutes.

In this study, calcium ions were used as the ions to be exchanged, and *tert*-butylphenol (TBP) was the hydrophobic solute. Several modified alumina packing materials were prepared by varying the polymerization period in the film-forming procedure, using a previously developed system. The ion exchange capacity was determined by monitoring the exit concentration of the calcium ions after injection began. In trying to increase the ion exchange capacity, sodium tetradecyl sulfate (STDS) was also adsorbed onto the modified aluminas. The ion exchange capacity was determined again for the STDS-treated alumina and compared to that without the STDS treatment.

1. Ion-Exchange Resins and Polymerized Films
   From SDS Admicelles on Alumina

A simple stoichiometric equation describing a reversible cation exchange reaction can be represented as

$$n\,R^-\,N^+ + M^{n+} \rightleftharpoons R_n M + n\,N^+$$

where R is the resin, $N^+$ is a monovalent ion, and $M^{n+}$ is the multivalent ion with n charges. The resin is usually a hardened polymer bead produced by the copolymerization of styrene and divinylbenzene under partial sulfonation of the phenyl groups. It is the sulfonated end groups that act as ion exchange sites. Other resins using alkylated ammonium functional groups serve for anion exchange.

The ion exchange capacity is defined as the total amount of ions exchangeable, based on a specific gegenion chosen as a reference. For the replacing ions, the ion exchange capacity based on a divalent ion will be half that on a monovalent ion. Therefore, an equivalent ion exchange capacity is defined as the capacity based on the equivalent moles of charge, without the ambiguity of different charges per ions. The units of this are milliequivalents (of charge) per gram (meq/g).

## B. Maximum Theoretical Ion Exchange Capacity of the Thin Film

We assume that bilayer coverage of SDS on alumina corresponds to the maximum attainable adsorption of SDS (4) and designate this adsorption as 100% coverage, even in the presence of styrene. Experimentally, the concentration of SDS is controlled to be between the onset of aggregate formation on the surface (the critical admicelle concentration, or CAC) and the formation of the first micelle in the bulk solution (the critical micelle concentration, or CMC). The polymerization reaction thus proceeds mainly in the admicelles, as discussed elsewhere (7). Under such circumstances, the coverage of SDS adsorbed on alumina may be far less than a complete bilayer coverage, depending on the pH of the system. The local bilayer aggregates, or admicelles, with the adsolubilizates can still be viewed as a sandwich structure. The upper and lower SDS layers surround the styrene interlayer. Below the CMC, however, they form in a patchwish manner and may not completely cover the surface at values of the pH too near the point of zero charge.

The stability of the SDS-polystyrene film complex is governed by the interaction of the sandwiched film with the adsorbent. If the solubility of the upper SDS layer is low enough not to be removed by the injected solution passing through the column, the ion exchange capacity of the filmed portion of the surface will be near the total SDS amount present. Experimentally, it was found that the upper SDS layer is slowly leached away by the injected aqueous solution. This problem can be addressed by substituting with sodium tetradecyl sulfate (STDS) as the surfactant.

The structure of STDS is similar to that of SDS, except that the carbon number is 14 instead of 12. Certain physical properties are, however, significantly different. For example, the Krafft temperature of the former is 48°C and that of SDS is 16°C. This difference results in the amount of STDS dissolved in an aqueous solution at room temperature to be very small compared to SDS. To dissolve STDS at concentration above the CMC, the temperature has to be above its Krafft point. Taking advantage of its low solubility in water at room temperature, it seems plausible to increase the ion exchange capacity of the filmed portion of the packing by applying STDS to the filmed alumina samples, hopefully forming a more stable upper layer of surfactant.

In applying STDS to the film surface, two possibilities exist. One is its adsorption on the bare alumina surface. The other is its adsorption on the polystyrene surface that has already been formed on a portion of the alumina. Since the film does completely cover the alumina in this study, the predominant locus of the adsorption depends not only on the adsorption isotherm for each pure adsorbent but also on the relative surface coverage. Data to determine the predominant location of the adsorbed STDS was not obtained in this study.

## C.  Experimental

### 1.  Materials

An intermediate surface area alumina ($100 \ m^2/g$) with a porous interior structure was employed in this study. The alumina powder was obtained by grinding 1/8-in. pellets (Alfa Products, No. 89372) into fine particles in a glass mortar and pestle. The resultant powder was then screened through a series of standard sieves, and particles between 45 and 75 μm and 75 and 105 μm were chosen. The solution of calcium ions was prepared by dissolving calcium chloride in deionized water having a conductivity less than 2 μmho. Chemicals used in forming the ultrathin film inside the pores of the alumina were the same as before (2,5–8). All other conditions and procedures were also unchanged, except for the reaction times. The column used here has a 6.2 mm inside diameter and a length of 25 cm. Calcium chloride was obtained from Sigma Chemical Company with a purity of 99%. The 4-*tert*-butylphenol was from Alfa Company, and of the same purity; it, however, was recrystallized from a hexane-ethanol-water mixture before use.

## 2. Procedures

The alumina powders coated with the ultrathin films were packed in a stainless steel column. Three or four burets were connected to a precision metering pump (Eldex Laboratories, model A-30S) for the accurate delivery of a fixed volume of solution from the burets to the column. A flow cell at the column outlet, with a pH electrode, was connected to a pH meter (Markson Science, model No. 6102). The solution, after passing through the flow cell, was collected in a fraction collector.

The SDS concentration in the fraction collector vials was determined by HPLC using a conductivity detector with a reversed phase C-18 packing material and a methanol-water mobil phase. Bulk conductivity of some samples was also measured with a separate conductivity detector.

For the measurement of sodium and calcium ion concentrations, an ion chromatography column (Wescan Instruments) was used. A cation column (No. 269-004) with appropriate eluants was used for measuring concentrations of sodium and calcium ions. A solution made with 1.25 mM nitric acid in water was prepared for the measurement of sodium. For the calcium, a mixture of ethylenediamine and tartaric acid in water at pH 4.5 was employed. The concentration of tartaric acid was 0.002 M; that of the ethylenediamine was 0.0017 M. In making all these solutions, deionized water was used. Before the HPLC measurements, these eluants were filtered several times by a glass filter under vacuum suction. Freshly prepared sodium chloride and calcium chloride solutions were used to establish the calibration curve for each ion. Good calibration curves were obtained for both.

The alumina powders with polymeric films were put in vials containing STDS to reach a partial coverage on the outside surface of the filmed alumina powders. These modified alumina samples were again packed in a column for the determination of ion exchange capacity, as described above. The Krafft temperature of STDS was measured as 48°C. To distinguish STDS adsorption from precipitation, the solution temperature has to be above the Krafft point. Temperatures of 50°C and 54°C were chosen for this purpose. An equilibration time of at least 1 week was allowed for the coating of the STDS onto the filmed alumina sample.

## D. Results and Discussion

The study made here was aimed at a demonstration of the separating power of the filmed alumina powders instead of a tracing of every component in the system. Figures 5–7 give the measured conductivity, pH, and exit SDS concentration for a sample prepared with a 1-hr polymerization time. For these studies, deionized water with a measured pH of 5.32 was injected. The conductivity decreases as the pore volumes injected increases beyond the appearance of the first peak. Since the bulk conductivity is an equivalent measurement of ionic species present, these ions may include sodium, hydrogen, and surfactant ions. In Fig. 6, the pH increases as the injected pore volumes increase. The decrease of hydrogen ion concentration is partially in agreement with the decreasing conductivity in Fig. 5. The conductivity decreases from the peak value of 0.5 mmho$^{-1}$ to approximately 0.1 mmho$^{-1}$, while the pH varies from 4.3 to 4.9.

Film stability or durability can be tested by inspecting the concentration of SDS eluted. It passes through a peak value of 400 μM then levels off gradually to 200 μM. A small square wave is observed at an injected volume of 50–100 ml, followed by a small peak covering 100–200 ml. This is due to the mass transfer resistance of the SDS inside the pore structure where the film was formed. The total amount of SDS eluted equals 74.7 μmol from a packing weight of 8.66 g of the filmed alumina. It is postulated that some SDS released from the alumina surface may be due to the leaching of the upper layer of the polymerized SDS aggregate. There is no spectroscopic evidence for this assertion, but it is consistent with the observed hydrophobic nature of the packing after washing.

These experiments were repeated, still using filmed alumina with a 1-hr reaction time, but at a lower flow rate. Results are shown in Fig. 8. The total injected volume was 300 ml. The pH increases after the peak wave at the initial stage of injection and approaches a final pH of 3.8 at the end of the injection. Since the powders were from different batches, the final pH values are different in Figs. 6 and 8. In Fig. 6, insufficient time was allowed to reach local equilibrium between the immobile phase and the mobile phase. In Fig. 8, the pH after 300 ml injection volume is still below

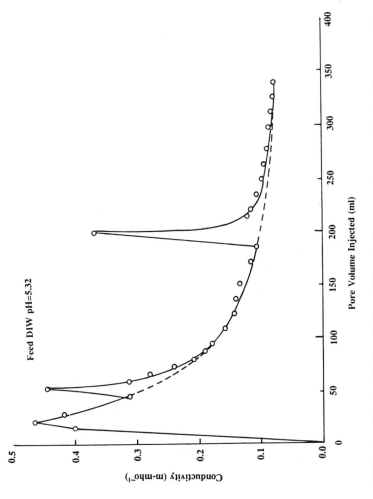

FIG. 5   Conductivity of the effluent from a column packed with a sample of the filmed alumina. The eluant is deionized water.

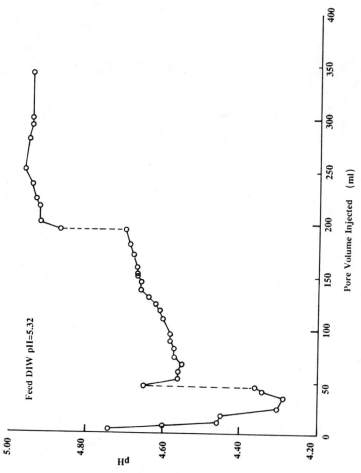

FIG. 6   Effluent pH from the same column as in Fig. 5.

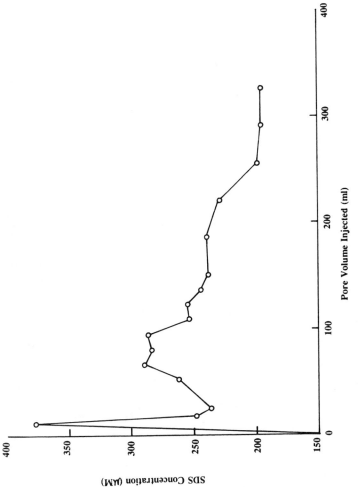

FIG. 7    Concentration of SDS in the effluent from the column in Fig. 5.

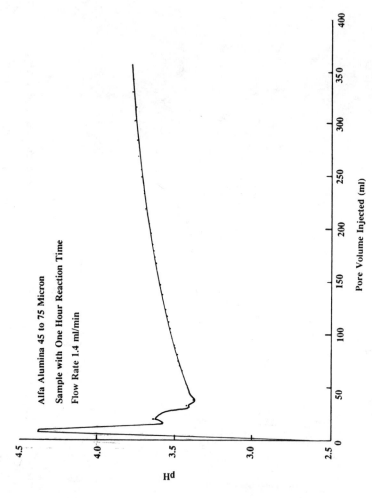

FIG. 8 Effect of a reduced eluant flow rate on the effluent pH for an identical bed.

4, which is an order of magnitude difference in hydrogen ion concentration from the feed acidity (pH 5.3).

To determine the ion exchange capacity of the alumina samples, a column packed with one of the aluminas was flooded with a 2 M sodium chloride solution for five pore volumes. Then one pore volume of deionized water was injected, to wash out the remaining sodium chloride solution. Finally, a 5000 $\mu$M calcium chloride solution was introduced as feed.

Figure 9 shows the exit concentration of the calcium for an unfilmed alumina sample. Although the pzc (point of zero charge) of the alumina is near 9, this sample shows a significant ion exchange capacity. A total capacity of 27 $\mu$mol/g of alumina was obtained after a 120-ml injection. Shown in Fig. 10 is a similar study for the filmed alumina, without styrene added during the polymerization step but keeping all other conditions the same. The ion exchange capacity is unchanged for this sample (neither sample was washed before being packed into the column).

Figure 11 shows the eluted concentration of sodium for a sample with a 45-min reaction time. The released or washed-out sodium concentration is high. The highest concentration is 0.6 M, which is much higher than that added at the beginning of the film-forming process. The filmed alumina was washed two or three times after the removal of the solution from the product by vacuum filtration, but even after this washing step a significant amount of sodium ions still remains.

Shown in Fig. 12 is the injected volume of $CaCl_2$ vs. the exit concentration of calcium ions. The feed concentration of calcium ions is 5000 $\mu$M. After a 30-ml injection, the calcium ion begins to appear in the effluent. At a 40-ml injection volume, the concentration goes up to 4800 $\mu$M, nearly the feed concentration. The ion exchange capacity of this material can thus be calculated to be 130 $\mu$eq per 10 g, or 0.013 meq/g. This is less than 5% of the maximum ion exchange capacity based on a stable polymerized film formed from SDS admicelles, assuming a surface area of 93 $m^2$/g for the alumina and a cross-sectional area of 2.5 $nm^2$ for the SDS headgroups. Although the theoretical maximum ion exchange capacity, counting the lower SDS layer, is 0.35 meq/g of filmed alumina—almost 13 times that of the unfilmed alumina—a very low ion exchange capacity material was obtained. That this capacity is actually less than that for the bare alumina (0.054 meq/g) is desirable for many applications of reversed phase packing materials. The principal reason for loss of ion exchange capacity may be

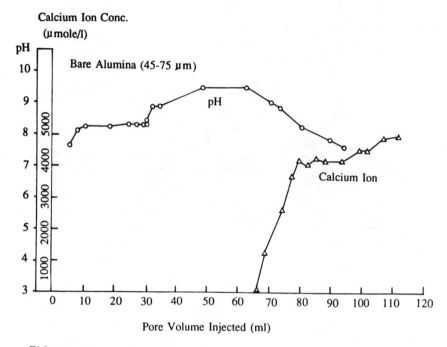

FIG. 9    Ion exchange capacity of a bed of unfilmed, unwashed alumina.

the nonwetted surface of the filmed alumina.    After the surfactant is leached from the film, it is hydrophobic in nature; this results in an incomplete exchange for cations, since some pores may not imbibe the bulk solution any longer.    It is not clear whether the 0.013 meq/g observed should be attributed to the film or to the unfilmed portion of the alumina surface.

Figure 13 shows the measured calcium concentration as a function of the injected volume of 5000 µM calcium chloride solution.    The packing sample had a 1-hr polymerization time and was subsequently treated with STDS.    The ion exchange capacity is almost doubled, compared to the sample not coated with STDS.    In addition to those ions exchanged by the polystyrene-SDS complex, the adsorbed STDS may also exchange sodium ions with calcium ions in the feed.    The additional contribution from the STDS adsorption may involve two mechanisms. One possibility is the adsorption of STDS onto the bare alumina

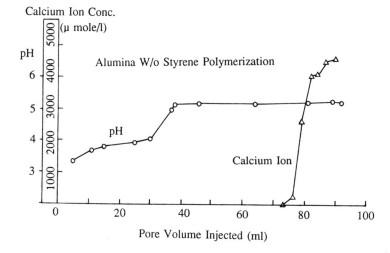

FIG. 10   Ion exchange capacity for a bed of alumina put through the film formation process, but without addition of styrene.

surface. The other is the adsorption of STDS to the poly-styrene films. Unfortunately, these data do not distinguish between the two.

In Fig. 14, the calcium ion concentration is plotted again as a function of pore volume injected for 5000 μM calcium chloride solution. The polymerization time for this sample was 45 min and the feed concentration of STDS was increased to 3000 μM. The ion exchange capacity of the sample was found to be 0.025 meq/g, a very slight improvement over the capacity of the material in Fig. 9.

In Fig. 15 the bed used in Fig. 14 is regenerated. The total amount of calcium ion regenerated from the bed corres-ponds to 0.025 meq/g, equal to the amount of calcium ion removed from the injected calcium solution. This result implies that the alumina/film structure is quite stable, at least in an aqueous environment. Finally, the bed was washed with deionized water to wash the sodium chloride solution from the pore space; then the calcium solution was reinjected. This time the ion exchange capacity was found to be 0.022 meq/g. This again implies that the film is very stable. In general, the

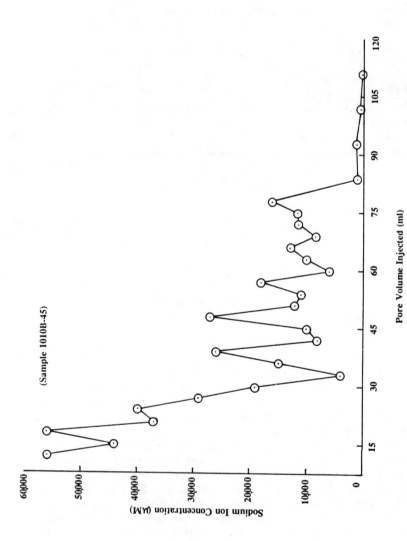

FIG. 11  Elution of sodium ions by deionized water from a sample of filmed alumina with a 45-min polymerization time.

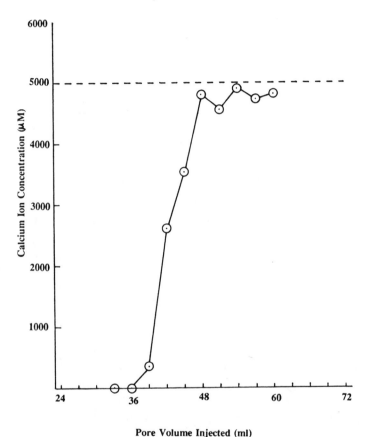

FIG. 12    Ion exchange capacity of a bed of filmed alumina.

STDS treatment increased the capacity of the filmed packing
by 75–80%.

Figure 16 shows the results obtained from using the packing directly after the polymerization step without washing
with deionized water.  This material has an ion exchange
capacity of 0.043 meq/g, higher than that of the washed
filmed samples, but still a little lower than that of the untreated
alumina.

For the study of the capacity of the packing to retain an
organic compound, a 2000 µM solution of TBP was first injected

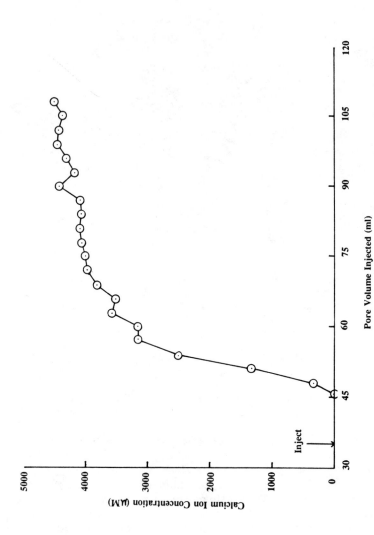

FIG. 13   Effect ot STDS treatment on the ion exchange capacity of the filmed alumina sample.

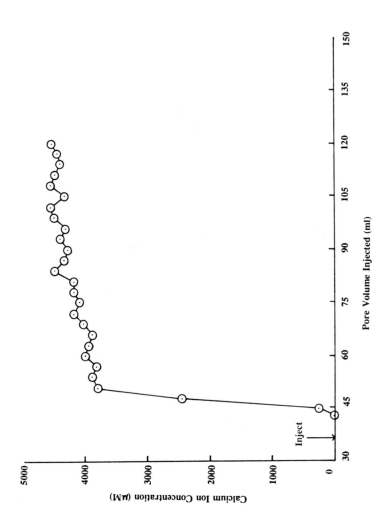

FIG. 14  Effect of increasing the STDS concentration during the STDS treatment of the filmed alumina.

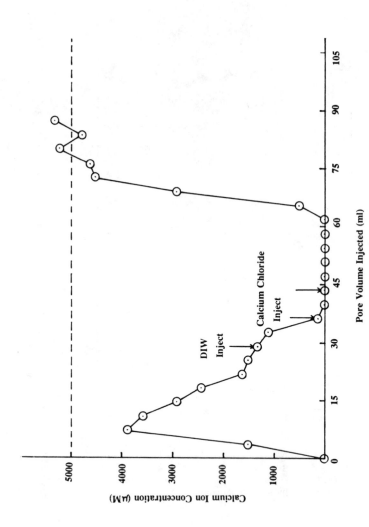

FIG. 15   Regeneration of the column shown in Fig. 14, followed by exami-
nation of the extent of retention of the ion exchange capacity.   This is a
test of the stability of polymerized film in the presence of an aqueous
solution.

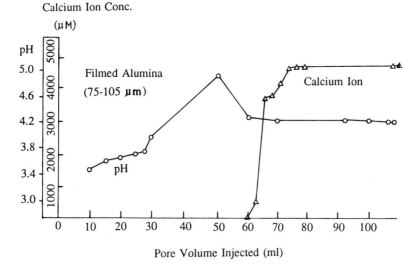

FIG. 16    Determination of the ion exchange capacity of an
unwashed sample of the filmed alumina without STDS treatment.

into a column packed with the unfilmed alumina.  As shown in
Figs. 17 and 18, the effluent concentration reached 1800 μM
after 15 pore volumes was injected.  Shown in Figs. 17 and 19
are the exit concentration of TBP and calcium ion for a filmed
alumina sample (No. 1216-60).  In these two figures, a prepared solution containing both TBP and calcium chloride was
injected instead of a single-component solution.  A simultaneous
removal of both TBP and calcium ions was obtained.  Too few
runs were made to speculate on why the breakthrough curve
for the calcium is so much sharper than the breakthrough
curve for the TBP.  The total ion exchange capacity is
0.034 meq/g, and the TBP removed is 0.124 mmol/g.  Calcium
breakthrough occurs earlier than the TBP breakthrough.
Though this ion exchange capacity is somewhat less than that
for the bare alumina, it is greater than that for the washed
filmed samples.  However, the removal of TBP by the filmed
alumina is excellent.  In fact, while the capacity for ion
exchange decreases by a factor of 4 with modification, the
capacity for organics has increased by orders of magnitude.
These results, summarized in Table 1, suggest that the filming
process described in this chapter can be used to create a

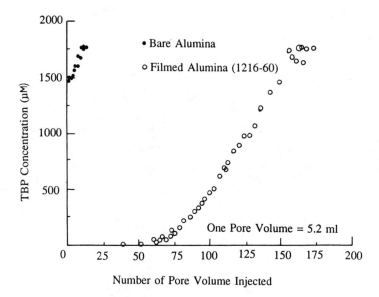

FIG. 17   Comparison of the removal of TBP from the eluant stream by unfilmed and filmed alumina packings.

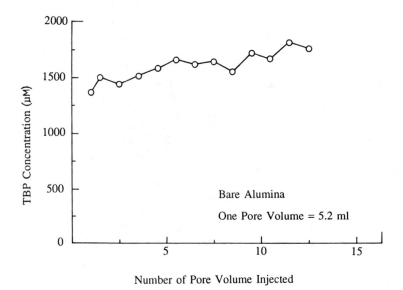

FIG. 18   Removal of TBP by the untreated, unfilmed alumina sample.

**TABLE 1** Summary of Calcium Ion and TBP Data for Alumina Samples Without Wash

| Packing type | Particle size (μm) | Packing wt. (g) | Calcium ion exchanged (μmol) | TBP removal (μmol) |
|---|---|---|---|---|
| Bare alumina | 75–105 | 9.528 | 260 | No inj. |
| Bare alumina | 75–105 | 8.900 | No inj. | 90 |
| W/o styrene polym. | 38–45 | 9.802 | 272 | No inj. |
| 1002A-60 film | 75–105 | 10.420 | 217.5 | No inj. |
| 1216-60 film | 75–105 | 10.460 | 174 | 1238 |

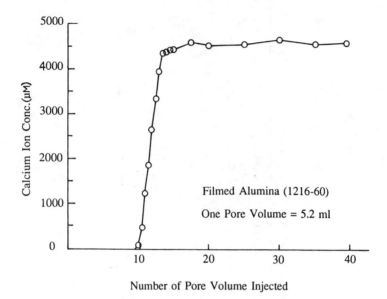

FIG. 19   Ion exchange capacity of the filmed alumina during the simultaneous removal of TBP shown in Fig. 17.

mixed mode packing with both ion exchange and reversed phase characteristics.

## V. CONCLUSIONS

1.   Surface modification of an inorganic packing using surfactants can significantly increase the affinity for organics while retaining ion exchange capacity.

2.   The upper SDS layer of the SDS-polystyrene complex is easily lost by washing with an aqueous solution. Specifically, the ion exchange capacity is higher for unwashed filmed samples than for washed ones, and SDS is observed to elute from beds packed with the filmed alumina.

3.   Due to the nonwetted nature of the filmed alumina surface, the ion exchange capacity for the filmed sample is less than that for bare alumina, although the SDS-polymer complex is expected to figure in the ion exchange behavior.

4.   The treatment of the filmed alumina samples by STDS increases the ion exchange capacity by nearly 80% over that of

untreated samples which were washed after the polymerization step.

5. A simultaneous removal of both organics and multivalent ions can be performed with the filmed alumina samples. The capacity for TBP removal is much greater than that for the cations. This suggests a possible route for producing reversed phase alumina with a retained low ion exchange capacity.

6. Packing made with the thin-film-forming process described in this chapter can be readily adjusted to obtain the desired balance between ion exchange capacity and hydrophobicity. The ion exchange capacity of the film could be increased even more by sulfonating the polystyrene, and the total capacity of the packing could be increased by beginning with a higher surface area alumina and by performing the polymerization step at a higher percentage coverage with admicelles.

## ACKNOWLEDGMENT

The recrystallized TBP used in these experiments was kindly provided by Mr. Bruce Roberts.

## REFERENCES

1.  C. Mira, Industrial Alumina Chemicals, ACS Monograph 184, 1986, Chap. 5.
2.  J. Wu, J. H. Harwell, and E. A. O'Rear, *Langmuir*, 3: 531 (1987).
3.  J. H. Harwell, J. Hoskins, R. S. Schechter, and W. H. Wade, *Langmuir*, 1. 251 (1985).
4.  D. Bitting and J. H. Harwell, *Langmuir*, 3: 531 (1987).
5.  J. D. Swalen, D. L. Allara, J. D. Andrade, E. A. Chandross, S. Garoff, J. Israelachvili, T. J. McCarthy, R. Murray, F. F. Pease, J. F. Rabolt, K. J. Wynne, and H. Yu, *Langmuir*, 3. 932 (1987).
6.  J. Wu, J. H. Harwell, E. A. O'Rear, and S. D. Christian, *AIChE J.*, 34: 1511 (1988).
7.  J. Wu, J. H. Harwell, and E. A. O'Rear, *J. Phys. Chem.*, 91: 623 (1987).
8.  J. Wu, J. H. Harwell, and E. A. O'Rear, *Colloid Surf.*, 26: 155 (1987).

# 9

# Surfactant-Enhanced Carbon Regeneration

D. LOWRY BLAKEBURN* and JOHN F. SCAMEHORN   Institute
for Applied Surfactant Research, University of Oklahoma, Norman
Oklahoma

---

*Present affiliation:  Conoco Inc., Ponca City, Oklahoma
Financial support for this work was provided by the Oklahoma
Mining and Mineral Resources Research Institute.   Hexcel
Corporation provided the surfactant used in these studies.

## SYNOPSIS

Surfactant-enhanced carbon regeneration (SECR) is a novel,
in-situ, low-energy method of regenerating spent adsorbent.
In SECR, a concentrated surfactant solution (regenerant solution) is passed through the adsorber after the carbon is
saturated with organic sorbate.  The organic sorbate desorbs
and is solubilized into the regenerant solution.  The result is
a very concentrated solution containing the original organic
for disposal or recovery of the solute and the surfactant.  The
residual surfactant on the carbon is then washed off with a
water flush step.  Since the surfactant used can be nontoxic
and biodegradable, this flush effluent can be directly diverted
to normal sewage systems.  Preliminary data are presented in
this chapter for a model phenolic solute that indicates the
feasibility of this new regeneration technique.

## I.  INTRODUCTION

Adsorption beds containing activated carbon are widely used to
remove dissolved organic pollutants from industrial wastewater
or gaseous emission streams.  This technology is also used in
recovery of organic products from gas or aqueous streams.

When the carbon bed becomes saturated with the organic
solute, the carbon must be regenerated or the adsorbed organic
removed from the carbon before the bed can be reused.  To
make the use of granular activated carbon economical at usage
rates of greater than 180 kg/day, regeneration of the carbon is

necessary (1). Regeneration of activated carbon is a major
factor in the cost effectiveness of the use of carbon (2).

The standard method of regeneration (1), thermal regeneration, involves removal of the carbon from the bed, transport of
the carbon to a hearth regeneration furnace where the adsorbed
organics are volatilized and carbonized, and loading of the bed
with regenerated carbon. This process is energy-intensive,
labor-intensive, and time-consuming. Further, the organic
adsorbate is not recovered, and up to 30% of the carbon may
be burned in the furnace.

Hot gas regeneration is an in situ regeneration method in
which a hot gas, such as steam or nitrogen, is passed through
the bed to desorb the adsorbate by a combination of purging
and desorption by heat-up effects (3,4). This is only effective when the adsorbate is highly volatile. Another in-situ
regeneration method is solvent regeneration (5,6), in which an
organic liquid solvent is passed through the bed to desorb
the adsorbate. This is also an energy-intensive process. When
the regeneration is complete, the solvent must be separated
from the organic adsorbate before reuse, and a thermal regeneration of the carbon must be performed to desorb the
residual volatile solvent. In biological regeneration (7),
another in-situ regeneration method, bacteria are introduced
into the bed to consume the adsorbed organic. Disadvantages
include the process being very slow, the organic not being
recovered, reduction of bed capacity from adsorption of some
of the products of the degradation, the need to induce desorption of the bacteria from the carbon when done, and the fre-
quent inability of the bacteria to ingest a mixture of organics.

## II. PRINCIPLES OF SURFACTANT-ENHANCED
##      CARBON REGENERATION

Surfactant-enhanced carbon regeneration (SECR) is a totally
new process that utilizes surfactants to remove adsorbed
organics from activated carbon in order to regenerate it for
reuse. In SECR, a concentrated surfactant solution is passed
through the spent carbon bed, as shown in Fig. 1. The
organic adsorbate desorbs and is solubilized into micelles in the
solution. Micelles are surfactant aggregates typically composed
of 50−150 surfactant molecules. The micelle has a hydrocarbon-
like interior into which organic molecules will dissolve or

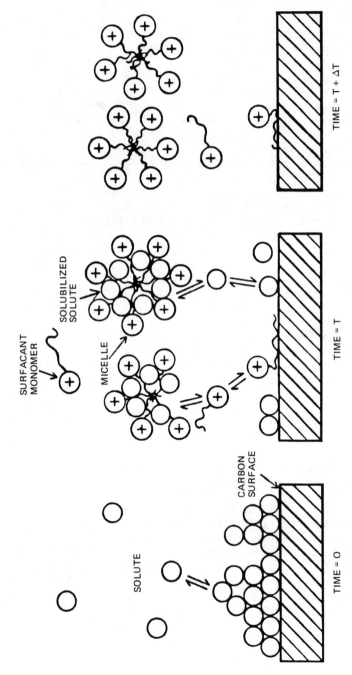

FIG. 1   Equilibria present at the various stages of a surfactant-enhanced carbon regeneration.

solubilize. A concentrated surfactant solution can contain
large concentrations of dissolved organics because of solubilization (8). Therefore, a small volume of the concentrated surfactant solution can potentially solubilize all of the adsorbed
organic in a spent carbon adsorption bed, resulting in a small
stream that has a high concentration of the organic adsorbate.

The surfactant can be removed from the regenerant solution exiting the adsorber by a precipitation process, as discussed in Chapter 12. This allows the surfactant to be recycled to the feed, which greatly improves the economics of
this process. The desorbed organic may now be present at
concentrations greater than saturation, resulting in a separate
organic-rich phase splitting out for direct reuse. The saturated
water could be recycled for use in the next regeneration cycle.
Alternatively, if the organic solute is very soluble in the water,
a very concentrated aqueous stream results from which the
organic can be separated by conventional techniques.

When the desorption/solubilization process is complete, the
carbon can contain some residual adsorbed surfactant. The
surfactant used can be chosen so that it has extremely low
toxicity and is biodegradable. Surfactants used in household

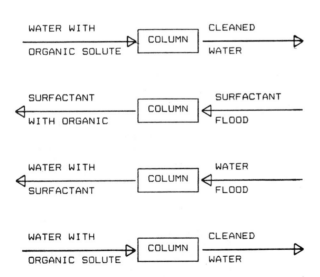

FIG. 2   Process strategy in surfactant-enhanced carbon
regeneration.

laundry applications are prime candidates. These environ-
mentally innocuous surfactants can then be flushed from the
carbon by water and the resultant stream simply diverted to
the normal sewage system, which is designed to handle these
kinds of detergents. No special recovery or clean-up is
needed. When done flushing, the carbon is ready for use for
liquid phase applications or needs to be dried for gas phase
applications. The process strategy for use of SECR is shown
in Fig. 2.

For solvent regeneration (5,6), solvents such as methanol
and acetone have been suggested. The carbon must be
cleansed of these residual solvents to a very high degree
before reuse of the carbon because of low allowable discharge
concentrations. For surfactants, on the other hand, the flush
probably does not need to be as complete as in solvent regeneration, since a higher surfactant concentration in discharge water
is more likely to be acceptable than for solvents such as
acetone or methanol. Also, the desorbed organic may be
similar enough to the regenerant solution in solvent regeneration that separation of the two from each other may not be
easy after regeneration, and may require an energy-intensive
technique such as distillation.


III.  EXPERIMENTAL


A.  Materials

The specific surfactant chosen for study was 1-hexadecylpyri-
dinium chloride monohydrate or cetylpyridinium chloride (CPC)
from HEXCEL Specialty Chemicals. It had a Krafft temperature
of 10.8°C (9). It was pharmaceutical grade as received, so it
was of high purity. Surface tension measurements on solutions
of CPC did not show a minimum, a classical test to detect im-
purities in surfactants (10,11). Also, analysis of solutions containing CPC with high-performance liquid chromatography
(HPLC) using UV and electrical conductivity detectors showed
a single, sharp peak—further indication of the purity of the
CPC.

The organic contaminant chosen for study was 4-tert-butyl-
phenol (TBP) from Aldrich Chemical Company. As received,
the TBP had a purity of 99%. The TBP was purified by recrystallization first from hexane and then from ethanol and
water. The melting point was found to be 99°C and very

sharp, indicating that the TBP was relatively free of impuri-
ties. Analysis of solutions containing TBP with HPLC using
an UV detector showed a single, sharp peak, also indicative of
the purity of the TBP.

The activated carbon used in this study was Calgon Filtra-
sorb 300, a granular, liquid phase carbon. It had a nitrogen
BET surface area of 950—1050 $m^2/g$, a mean particle diameter
of 1.5—1.7 mm, and an effective size of 0.8—0.9 mm, as
reported by the manufacturer (12). The carbon was boiled
rapidly for 4—6 hr and then rinsed several times with deionized
water. The remaining salts were subsequently desorbed from
the carbon surface by rinsing the carbon three times, this
being repeated every 2 days, for 15—20 days, until no salts
were detected in the rinse water. The carbon was then dried
for 1 week at 100—120°C. Due to the grinding action on the
carbon when it was boiled in water, the final mean diameter of
the cleaned carbon particles was 1.0—1.2 mm.

## B. Methods

### 1. Regeneration and Flushing of Adsorber Bed

The carbon adsorption bed was a 480.0-ml glass column, 25 mm
diameter by 100 mm length, with a maximum pressure rating of
690 kPa, manufactured by Rainin. An adjustable plunger was
used to ensure that no excess void volume existed in the bed.
A water jacket was used to maintain the carbon bed at 30.0°C.
A single head pump with a maximum flow rate of 82 ml/min and
a maximum pressure of 690 kPa, manufactured by Cole-Palmer,
was used in this study. For experimental adsorption columns,
the ratio of the column diameter to the packing diameter
should be at least 20:1 (3,13). A minimum of 25-mm-diameter
column has been suggested for granular carbon adsorption bed
studies (12). The adsorption bed used here satisfied these
criterion.

In the preparation of the carbon to be used in the actual
regeneration, a 205-g carbon sample was boiled in water to
completely wet the pores, after which the excess water was
decanted. The water/carbon ratio was known at the end of
the boiling/decanting step. A known amount of water and
30.0 g of TBP was added to the wet carbon and the mixture
was heated to 80°C to dissolve the TBP in the water. The
mixture was then placed in the carbon adsorption column. The
TBP solution was recirculated through the carbon adsorption

bed at 30.0°C until equilibrium was reached, as shown by the
TBP concentration in the column effluent. As a result, the
equilibrium concentration of TBP and the loading on the carbon
was known at the start of the regeneration. In this bed, one
pore volume (void volume in the bed) equaled 309 ml.

In the regeneration step, a 0.4 M solution of CPC was
pumped through the column at a flow rate of 1−2 ml/min in a
downflow configuration. Therefore, the average residence
time of the regenerant solution was about 200 min. Samples
of the effluent from the column were collected and the CPC
and TBP concentrations measured using HPLC with conductivity
and UV detection, respectively.

During the water flush step, water was pumped through
the column in a downflow configuration at a flow rate of 1−2
ml/min and samples were also taken on a periodic basis, and
the CPC and TBP concentrations in these samples were
measured as in the regeneration step.

## 2.   Miscellaneous Measurements

The CMC of CPC solutions containing a given concentration of
TBP was determined from the break in the surface tension
curve as a function of CPC concentration (14). The du Nouy
ring method was used to measure the surface tension of the
surfactant solutions at 30.0°C, taking the necessary precautions for accurate measurements (15).

In the measurement of static adsorption isotherms of TBP
on carbon, a carbon sample of known weight was placed in a
1000-ml boiling flask with about 300 ml of water. By varying
the weight of the carbon sample, the solvent/solid ratio was
varied from 10 to 5000 ml/g. The carbon and water mixture
was boiled rapidly to cause the water to wet the carbon pores.
The carbon and water mix was then cooled, the water drained
off, and the carbon rinsed several times to remove any salts.
A known amount of TBP was then added along with 100 ml of
water to the wet carbon in a 120-ml bottle. The sample was
sealed and maintained at 80°C for approximately 30 min to drive
the TBP into solution. The sample was then maintained at
30.0°C for approximately 2 weeks. The TBP equilibrium concentration was measured using HPLC with UV detector and the
adsorption of TBP on the carbon calculated from a material
balance. Other experimenters (16) have ground the carbon
into a fine powder to maximize the surface area of the carbon
and to reach equilibrium in a short time. This can change

the physical surface of the carbon, so the cleaned granular
carbon was used here instead of the powder.

In the measurement of the static adsorption isotherm of
CPC on carbon, a carbon sample of known weight was placed
in a 100-ml bottle and a known volume of a CPC solution of
known concentration was added. The presence of the CPC
caused the aqueous solution to wet the pores of the carbon.
To ensure complete wetting, a vacuum was pulled on the
slurry for 2 hr, during which time evaporation of water was
negligible. By varying the weight of the carbon sample, the
solvent/solid ratio was varied between 5 and 50 ml/g. The
sample was equilibrated at 30.0°C for approximately 72 hr and
the CPC concentration measured by HPLC with an UV detector.

## IV. MECHANISTIC CONSIDERATIONS IN SURFACTANT-ENHANCED CARBON REGENERATION

Prior to discussing results from application of SECR in an adsorption bed, the individual physical steps involved in the
process will be discussed.

### A. Organic Contaminant on Carbon Bed

When most of the activated carbon in the bed is saturated with
the organic adsorbate, the organic solute concentration in the
effluent from the bed begins to increase dramatically (1). This
is called breakthrough and the bed must be regenerated before
reuse. For a properly designed bed, the adsorption capacity
at breakthrough is within a few percent of that if the entire
bed were at equilibrium with the feed solution (i.e., the vast
majority of the bed is saturated with the organic). The saturation adsorption level increases with feed concentration.

The adsorption isotherm for TBP on carbon at 30.0°C is
shown in Fig. 3. It may be described by a Freundlich adsorption isotherm (16,17):

$$Q = 86C^{0.19} \tag{1}$$

where Q is the adsorption in mg TBP/g carbon and C is the
equilibrium concentration of TBP in solution in micromolar.

This adsorption curve was measured at a constant pH. All
of the water used to make the samples and the final equilibrated
samples had pH of 4–5. The solvent/solid ratio apparently had

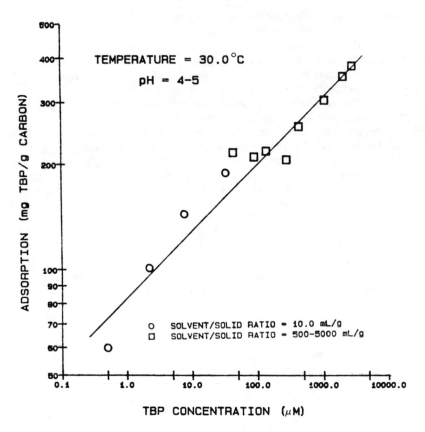

FIG. 3   Equilibrium adsorption isotherm of TBP on activated carbon.

no effect on the TBP adsorption isotherm over the more than two orders of magnitude range shown in Fig. 3.   This indicates that all parameters significantly affecting the adsorption were being either measured or held constant.

## B.   Surfactant Monomer Concentration

When a surfactant is dissolved in water below a concentration called the critical micelle concentration (CMC), no micelles are present; all the surfactant is present as unassociated species or monomer.   At total surfactant concentrations above the CMC, some of the surfactant is present as monomer and some as micelles. For ionic surfactants with an added electrolyte, the surfactant monomer concentration is less than or equal to the CMC at total surfactant concentrations above the CMC (18). The CMC may be affected by any additives present in solution, such as TBP.

The values of the CMC of aqueous solutions of CPC are shown in Fig. 4 as a function of the TBP concentration in solution. A CPC concentration of 0.4 M was used in the actual regeneration step studied. All of the CMC data shown in Fig. 4 is below 0.001 M. Therefore, under the conditions of interest, greater than 99% of the surfactant is in micellar form and little is wasted as monomer.

## C. Solubilization of Organic Contaminant in Surfactant

Solubilization is the mechanism by which the aqueous surfactant solution can dissolve very high concentrations of the adsorbed solute during the regeneration step. It is the ability of the micelles to solubilize large quantities of organic solute which make SECR feasible.

In an aqueous micellar solution, the solute can either be solubilized in the micelles or can be present as unsolubilized in bulk solution. The relative concentrations of solubilized and unsolubilized solute can be represented as a distribution coefficient:

$$K = \frac{C_s}{C_u C_m} \tag{2}$$

where K is the distribution coefficient, $C_s$ is the concentration of solubilized organic, $C_u$ is the concentration of unsolubilized organic, and $C_m$ is the concentration of surfactant in micellar form. K is independent of concentration if Henry's law applies to the solubilization of organic in surfactant micelles (10). For TBP solubilizing in solutions of CPC, $K = 1350 \ M^{-1}$ (9). Therefore, in the regenerant solution containing 0.4 M CPC, the concentration of solubilized TBP/concentration of unsolubilized TBP is 540, indicating that 99.8% of the TBP in solution is solubilized.

## D. Adsorption of Surfactant on Carbon Bed

As the regenerant solution passes through the carbon bed, the organic sorbate is desorbing and solubilizing in the solution.

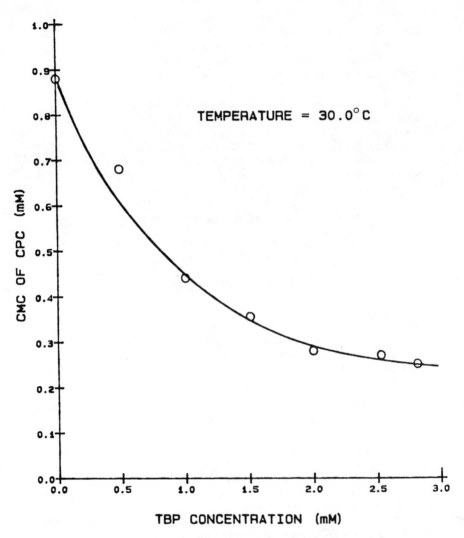

FIG. 4    Effect of TBP concentration on the CMC of CPC
solutions.

However, surfactant is simultaneously adsorbing on the carbon. This has several disadvantages. Obviously this is a waste of surfactant. Also, this adsorbed surfactant must be removed in the flushing step. If no surfactant were adsorbed, there would be no need for a flushing step. Another problem is that the adsorbed surfactant can form micelle-like aggregates in a monolayer-type structure on the surface. The organic solute can coadsorb with this surfactant due to the affinity for the organic-like interior of the monolayer structure. This phenomena is similar to the solubilization in micelles and is called adsolubilization, as discussed in detail in Chapter 7. Adsolubilized solute may be difficult to desorb and remove during the regeneration step.

The adsorption of CPC on carbon was measured at 30.0°C over a wide range of CPC concentrations as shown in Fig. 5. The pH of all the final equilibrated samples was 4−5. The solvent/solid ratio, varied from 5 to 50 ml/g, did not visibly affect the adsorption of CPC on carbon.

## E.  Other Properties of the Surfactant Solution

If enough of an organic solute is present, the surfactant solution may phase-separate or form macroemulsions, which are very viscous. The maximum solubility of TBP in a CPC solution, without phase separation, occurred in about 0.4 M CPC (9). Solutions of CPC at higher concentrations than 0.4 M were increasingly viscous, even without the TBP present. Therefore, the CPC concentration in the regenerant solution was limited to 0.4 M.

Since some of the surfactant will be emitted to the environment or to sewage systems during the flush after regeneration, the environmental impact is of concern. About 99% of the original surfactant hydrocarbon chain would be gone within 2 weeks after entering the environment due to bacterial decomposition (19). The waters of most industrialized countries contain an excess of anionic surfactants. A cationic surfactant will form an electrically neutral salt with an anionic surfactant. The result will be a very slightly soluble molecule that is biodegradable, at least from the cationic hydrocarbon end (20). Overall, CPC is not expected to pose a significant environmental risk in wastewater treatment (21).

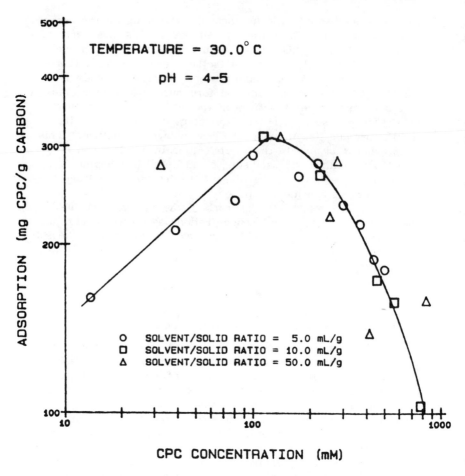

FIG. 5    Equilibrium adsorption isotherm of CPC on activated carbon.

F.  Selection of System Studied

The three major classes of surfactants were considered for use
in SECR:  nonionic, cationic, and anionic.  The ethoxylated
nonionic surfactants used in screening studies formed macroemulsions with the TBP in the system, making them unsatis-
factory.  If the surfactant is restricted to a single component,
cationic surfactants are superior to anionic surfactants for
several reasons.  An anionic surfactant is limited to a shorter
hydrocarbon chain length than a cationic surfactant if the
system is constrained to be above the Krafft temperature of
the surfactant (9).  Below the Krafft temperature, the surfactant precipitates from solution.  These longer alkyl chain
lengths allow cationic surfactants to exhibit higher solubilization
capacities and a lower CMC than anionic surfactants (9).  Therefore, cationic surfactants are the preferred surfactant type.
Some of these same advantages can be obtained for anionic surfactants by using surfactant mixtures (22), but a single surfactant type is appropriate for the first demonstration of this
new process.
    TBP is representative of phenolic compounds, which repre-
sent important environmental problems, mainly as dissolved
species in water.  TBP is used in industry as an intermediate
in the manufacture of varnish and lacquer resins, it is a soap
oxidant, it is an ingredient in the deemulsifiers for oil field
use, and it is a motor oil additive (23).  TBP represents an
environmental problem itself as well as representing phenolics
as a class.
    The Filtrasorb 300 is a standard liquid phase activated
carbon.


V.  TEST OF SURFACTANT-ENHANCED CARBON
    REGENERATION IN AN ADSORPTION COLUMN

A.  Regeneration

The concentration of TBP in the solution in equilibrium with
the carbon bed was 13 µM prior to regeneration.  The loading
on the simulated spent carbon was 0.146 g TBP/g dry carbon.
This represents a spent carbon adsorber which has been used
to remove TBP from a feed solution containing 13 µM TBP.

Under these conditions, 32,000 pore volumes of wastewater could have been cleaned up by the bed.

The concentrations of TBP and of CPC in the column effluent during regeneration are shown in Figs. 6 and 7, respectively. The TBP concentrations are converted into fractional recoveries in Fig. 8. From Fig. 6, the TBP concentration in the bed effluent was extremely high, reaching 0.122 M when it first appeared. It then declined as the regeneration step proceeded. From Fig. 8, after 10 pore volumes of regenerant solution had emitted from the column, 48% of the TBP had been removed. The ratio of the volume of wastewater that the carbon could have treated to that of the regenerant solution at 10 pore volumes of regenerant solution is 3300. After 160 pore volumes, 79.4% of the TBP had been removed. The ratio of the volume of wastewater that the carbon could have treated to that of the regenerant solution at 160 pore volumes of regenerant is 205. Therefore, 48% of the TBP had been regenerated from the column when only 0.03% of the volume of wastewater that could be originally treated is used in the regeneration. The TBP in the regenerant solution at this point is 20.4 mM or 1574 times as concentrated as in the original wastewater treated. Similarly, 79.4% of the TBP had been regenerated from the column when only 0.487% of the volume of wastewater which was originally treated was used in the regeneration. The TBP in the regenerant solution at this point is 2.11 mM or 162 times as concentrated as in the original stream treated. However, the TBP concentration in the regenerant solution is very dilute at large pore volumes, indicating that the last 20% is very difficult to desorb.

From Fig. 7, the CPC was emitted after an average of approximately 1.5 pore volumes. If none had adsorbed, it would have come out at 1 pore volume. From a material balance, 38.1 g of CPC adsorbed, compared to 42.0 g predicted from the static adsorption isotherm shown in Fig. 5. Considering the complicated situation during the regeneration with some TBP still adsorbed, this indicates that static surfactant adsorption isotherms can approximate the adsorption during the regeneration process.

This difficulty in removing the last 20% of adsorbate may be due to this fraction of adsorbate being chemisorbed instead of physically adsorbed. The organic that is chemisorbed on the carbon may be very difficult to remove. However, in a recurring regeneration, this may not be a problem, since those adsorption sites corresponding to chemisorption may remain filled

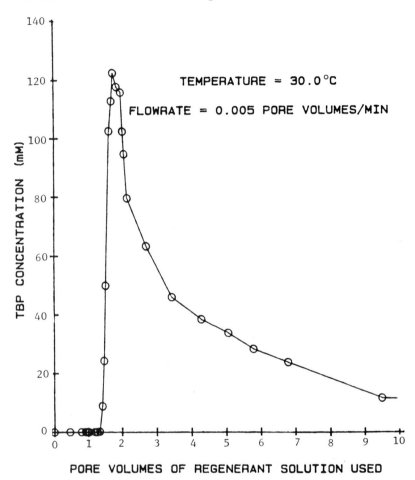

FIG. 6   Concentration of TBP in effluent from regeneration.

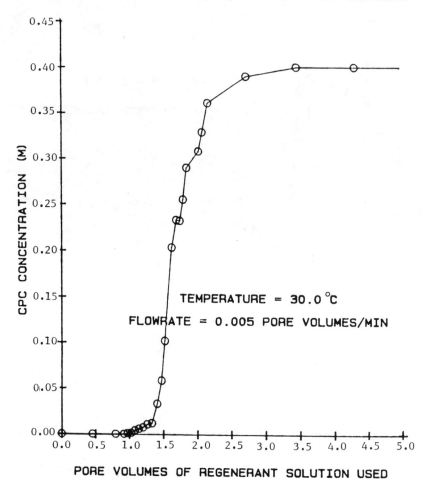

FIG. 7   Concentration of CPC in effluent from regeneration.

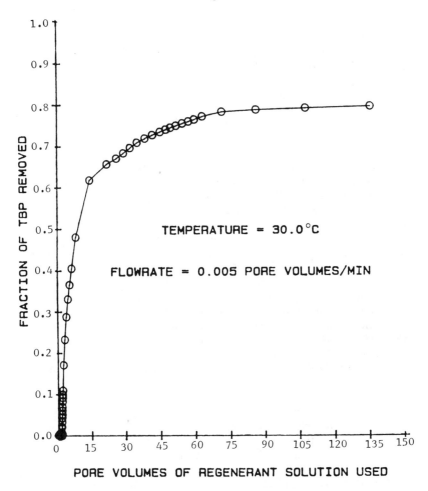

FIG. 8    Fractional removal of TBP during regeneration.

and not take further part in adsorption processes.  If this is
the case, subsequent regenerations would recover essentially
all of the TBP in a few pore volumes (less than 200).  This
hypothesis has been qualitatively proven in subsequent work in
which the recovery of solute approached 80% during the first
regeneration and approached 100% during subsequent regenerations.

Taking a very simplified view of the regeneration, let us
assume that the regenerant solution solubilizes desorbed TBP
at a level such that the unsolubilized TBP concentration in the
regenerant solution is equal to the TBP concentration in
equilibrium with the spent carbon prior to regeneration (13 $\mu$M).
In other words, the 13 $\mu$M unassociated TBP is assumed to be
in equilibrium with the adsorbed TBP and the micelles in the
regenerant solution simultaneously.  If this is the case, the
previously discussed solubilization equilibrium constant can be
used to calculate that 533 pore volumes of regenerant solution
can be used to remove all of the TBP from the bed.  The TBP
concentration in the regenerant solution would be 800 $\mu$M
throughout the regeneration.  This idealized case will be referred to as the equilibrium solubilization case.  The ratio of
the volume of the wastewater that could have been treated to
that of the regenerant solution when regeneration is complete
could be 61.5.  The TBP in the regenerant solution at this
point would be 0.8 mM or 61.5 times as concentrated as in the
original stream treated.

In the removal of the first 80% of the TBP, the actual regeneration performed better (required less regenerant solution)
than the equilibrium solubilization case.  This is due to the
complex chromatographic interaction as the regenerant solution
passes through the column (a more detailed discussion of
chromatographic effect during desorption in fixed beds is given
in Chapter 7).  Even if removal of some residual portion of the
adsorbate from the carbon is very difficult, this might reduce
the efficiency of the operation very little if appropriate process
configurations are used.  The flow of the wastewater through
the regenerated bed can be in opposite direction to that of the
regenerant solution and flush solution, as shown in Fig. 2.
The water stream leaving the adsorber would then be exposed
to the freshest carbon (that which was regenerated most efficiently) just before leaving the bed.  The net effect could be
very little reduction in final adsorption capacity after this
regeneration process.

## B.  Water Flush

After approximately 160 pore volumes of regenerant solution had
been passed through the bed, the flush step was initiated.
The concentrations of CPC and TBP in the flush solution are
shown in Figs. 9 and 10, respectively.  The fractional removal
of CPC during the flush is shown in Fig. 11.  Both the effluent CPC and TBP concentrations decrease rapidly as the flush
step proceeds.  After the use of 10 pore volumes of flush solution, 26% of the residual CPC had desorbed; after 160 pore
volumes, 51% of the residual CPC had desorbed.  After 1460
pore volumes, the effluent still contained 40 μM CPC.  Although
the amount of residual TBP adsorbed on the carbon after the
regeneration is unknown, the concentration of TBP in the effluent has a similar profile to that of the CPC (Figs. 9 and 10).
This would imply that adsolubilization may be a major mechanism
of entrapment of TBP after regeneration and that the TBP desorbs roughly in proportion to the CPC.

The conclusion concerning the flushing step is similar to
that concerning the regeneration step, namely, that the removal
of the bulk of the residual of the adsorbed material can be
achieved with a relatively small volume of solution.  However,
removal of the last portion of the residual material can require
large quantities of flushing water.

The fraction of the surfactant being chemisorbed, as op-
posed to physically adsorbed, may be the cause of this effect,
as was speculated for the analogous step in the regeneration.
If true, the removal of essentially all of the surfactant adsorbed
during the regeneration during the subsequent flushing may be
possible for cycles after the first one, which was studied here.
Also, other surfactants may not adsorb as much as CPC on the
carbon, minimizing the importance of the flush step.  Finally,
it might prove worthwhile to pretreat the carbon with an agent
that would block any chemisorption sites from the surfactant,
allowing 100% recovery of the process surfactant in a few pore
volumes of flush water.  Current work is exploring these
possibilities.

## C.  Practicality of Regeneration

The results here indicate that the basic premise behind this
new method of regenerating activated carbon is valid.  Large
concentration increases in the organic adsorbate in the regenerant solution compared to that in the original wastewater

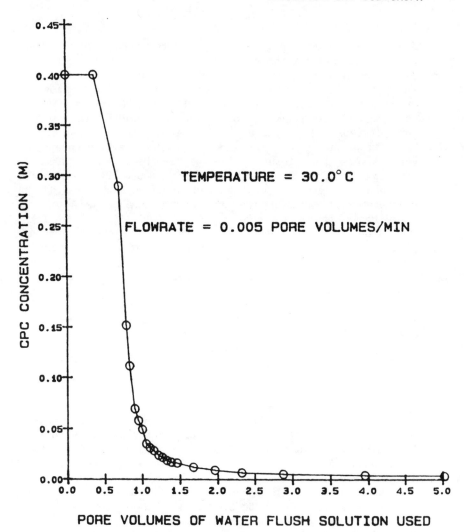

FIG. 9   Concentration of CPC in effluent from water flush.

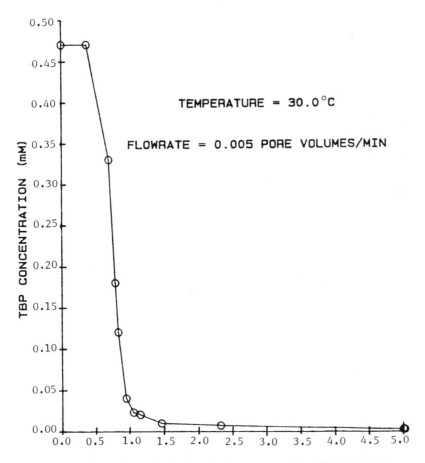

PORE VOLUMES OF WATER FLUSH SOLUTION USED

FIG. 10    Concentration of TBP in effluent from water flush.

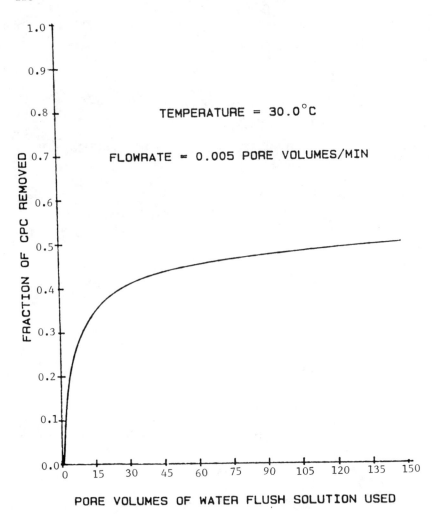

FIG. 11   Fractional removal of CPC during water flush.

treated were observed and low volumes of regenerant solution were generated compared to the volume of original solution treated. The kinetics of removal of the bulk of the adsorbate were favorable.

Standard engineering design strategies of using counter-current flows for the wastewater and the regenerant solution and/or having a small polishing bed downstream of the main bed to meet required purities could improve the economics of this process even further.

In summary, SECR shows great promise as a universal, in-situ, nondestructive carbon regeneration technique that may find wide use in future applications.

## REFERENCES

1.  R. A. Hutchins, in *Handbook of Separation Techniques for Chemical Engineers* (P. A. Schweitzer, ed.), McGraw-Hill, New York, 1979, Sect. 1.13.
2.  R. W. Soffel. in *Kirk-Othmer Encyclopedia of Chemical Technology*, 3rd ed., Vol. 4, Wiley, New York, 1983, p. 563.
3.  J. L. Kovach, in *Handbook of Separation Techniques for Chemical Engineers* (P. A. Schweitzer, ed.), McGraw-Hill, New York, 1979, Sect. 3.1.
4.  J. F. Scamehorn, *Ind. Eng. Chem. Process Des. Dev.*, *18*: 210 (1979).
5.  T. Sutlkno and K. J. Himmelstein, *Ind. Eng. Chem. Fund.*, *22*: 420 (1983).
6.  P. C. Wankat and L. C. Partin, *Ind. Eng. Chem. Process Des. Dev.*, *19*: 446 (1980).
7.  W. A. Chudyk and V. L. Snoeyink, *Environ. Sci. Technol.*, *18*: 1 (1984).
8.  P. H. Elworthy, A. T. Florence, and C. B. MacFarlane, *Solubilization by Surface Active Agents*, Chapman and Hall, London, 1968, Chap. 2.
9.  R. O. Dunn, J. F. Scamehorn, and S. D. Christian, *Sep. Sci. Technol.*, *20*: 257 (1985).
10. J. H. Clint, *J. Chem. Soc., Faraday Trans. I, 71*: 1327 (1975).
11. M. J. Rosen, *Surfactants and Interfacial Phenomena*, Wiley, New York, 1978, Chap. 6.
12. Calgon Corporation, *Adsorption Handbook*, Activated Carbon Division, Calgon Corporation, Pittsburgh, Pa.

13.   C. G. Hill, *An Introduction to Chemical Engineering Kinetics and Reactor Design*, Wiley, New York, 1977, Chap. 6.

14.   P. Mukerjee and K. J. Mysels, *Critical Micelle Concentrations of Aqueous Surfactant Systems*, National Bureau of Standards, Washington, D.C., 1971.

15.   K. Lukenheimer and K. D. Wantke, *Colloid Polym. Sci.*, *259*: 354 (1981).

16.   R. A. Dobbs and J. M. Cohen, *Carbon Adsorption Isotherms for Toxic Organics*, EPA, Cincinnati, 1980.

17.   A. W. Adamson, *Physical Chemistry of Surfaces*, 4th ed., Wiley, New York, 1982.

18.   K. J. Mysels, *J. Colloid Sci.*, *10*: 507 (1955).

19.   R. D. Swisher, *Solution Behavior of Surfactants: Theoretical and Applied* (K. L. Mittal, ed.), Plenum Press, New York, 1982, p. 149.

20.   L. H. Huber, *J. Am. Oil Chem. Soc.*, *61*: 377 (1984).

21.   R. S. Boethling, *Environmental Fate and Toxicity in Wastewater Treatment of Quaternary Ammonium Surfactants*, EPA, Washington, D.C., 1984.

22.   J. F. Scamehorn, in *Phenomena in Mixed Surfactant Systems* (J. F. Scamehorn, ed.), ACS Books, Washington, D.C., 1986, Chap. 1.

23.   M. Windholz (ed.), *Merck Index*, 9th ed., Merck and Company, Rahway, N.J., 1976.

# IV
## SEPARATIONS BASED ON FOAMS

# 10
# Adsorptive Bubble Separation Processes

**THOMAS E. CARLESON**  Department of Chemical Engineering,
University of Idaho, Moscow, Idaho

SYNOPSIS

The factors affecting the adsorptive bubble separation of
materials are discussed and the variety of applications sur-
veyed. Foam separation or fractionation, bubble fractionation,
and solvent sublation techniques are summarized. Examples of
the separation of low concentrations of biological materials,
cationic and anionic species, and organics from their solutions
are presented. The factors of inherent surface activity or
complexation of the species to be separated with a surface-
active substance (surfactant), flow rates of the phases, concentration of species, pH, and interfacial properties of the two
phases are discussed relative to their effect on the separation.


I.  INTRODUCTION

Adsorptive bubble separation is a phrase coined by Lemlich
(1,2) to describe a process where a species is adsorbed at an
interface between a dispersed phase (bubbles) and a continuous
phase. The dispersed phase with the adsorbed substance is
subsequently collected. This definition encompasses two main
processes (2–7). One of these is foam separation, which includes foam fractionation and flotation (ore flotation, macro-
and microflotation, precipitate flotation, ion flotation, molecular
flotation, and adsorbing colloid flotation). Flotation involves
the removal of particulate by frothing, whereas foam fractionation involves the separation of soluble species by foams.
    The other main division of adsorptive bubble separation is
nonfoaming adsorptive bubble separation. This includes bubble
fractionation (use of a gas for adsorption and deposition of a
concentrated solute at the top of the main liquid) and solvent
sublation (use of another liquid phase to adsorb and concentrate
the solute at the top of the main liquid phase). In most of
these processes, the dispersed phase consists of air bubbles
and the continuous phase, generally water, contains the species
being adsorbed, although oil has been used as the dispersed
species in ore flotation.
    Since Chapter 11 deals with froth flotation (including ore
flotation, precipitate flotation, and other particulate phase
separation processes), this chapter will deal only with separation of molecular species or soluble species by adsorptive bubble
separation (foam fractionation, solvent sublation, bubble
fractionation).

In order for adsorption between phases to occur, the adsorbed species must be surface-active, i.e., tend to migrate to and concentrate at the interface. For a continuous water phase, the species must have hydrophobic and hydrophilic molecular groups. Hydrophobic groups are organic molecular groups and consist of long-chain hydrocarbon groups of fatty acids and alcohols, or cyclic groups attached to alkyl groups such as alkyl benzene in sodium alkyl benzene sulfonate and the propylated naphthalene group in the sodium salt of propylated napthalenesulfonic acid. The hydrophilic groups include the charged portion of the molecule such as the sulfate, amine, or carboxyl group. They can have a positive charge as in cetyltrimethylammonium bromide (a cationic surfactant), a negative charge as in sodium dodecylbenzene sulfonate (an anionic surfactant), or be neutral as in sorbitan monopalmitate (a nonionic surfactant).

The adsorbing species may be formed by the addition of a surface-active agent (a "collector") to a solution of the species to be removed. The two must react to form a complex that is surface-active. For instance, many heavy metals can form complexes with anionic surface-active agents such as sodium dodecyl sulfonate (4). The subsequent complex can be foamed off. A number of factors affect the formation and adsorption of the complex at the bubble surface (8). These include the concentrations of the metal ion, the surfactant, and the complex; the pH; ionic strength; temperature; and the concentration of competing surface-active species. This same mechanism is used to separate particles by adsorption of a surface active agent on the particle to render it hydrophobic. Electrical interactions between the particle and the surface-active molecule are important (9).

As the dispersed or bubble phase travels through the continuous phase, mass transfer of the surface-active solute occurs between the two phases. The driving force for transfer is the difference in concentration between the surface phase concentration corresponding to equilibrium with the bulk phase and the actual surface phase concentration. In many cases, adsorption equilibrium can be determined from an appropriate adsorption isotherm expression, such as the Langmuir or Frumkin isotherm. The continuous phase concentration of surfactant must be low (not more than a few hundred parts per million for an aqueous continuous phase). For elevated surfactant concentrations, micelles (clumps of surfactant molecules in the continuous phase) form which increase the solubility of the adsorbing species in the continuous phase relative to the interface.

If the bubbles collapse before exiting the continuous phase, there will be a region at the top of the continuous phase of a higher concentration of the adsorbed species. This is called bubble separation for air bubbles as the dispersed phase and solvent sublation for another immiscible liquid as the dispersed phase (3,4). As the bubbles of air exit the continuous phase, the liquid lamella between the bubbles drains. It the liquid contains the appropriate substances, the liquid will not drain completely and a foam will issue from the continuous phase.

Foam stability is a complicated subject and many papers have been dedicated to its study. From thick-walled (wet) foams, draining is by gravity or by suction from the Plateau borders (regions at the intersection of bubbles where there is high curvature and consequently high-pressure gradients). As the flow proceeds, adsorbed surfactant can be compressed and result in a surface force opposing the flow direction. This is the phenomenon of Gibb's elasticity that can stabilize foams, although in some cases transfer of a surface-active species can result in rapid foam thinning and rupture. As films thin to a few hundred nanometers, due to the approaching interfaces, other forces come into play.

Electrical charges on the interfaces can affect the stability and result in an effective pressure normal to the thinning surfaces. This force has been termed the disjoining pressure by Derjaguin (10) and is due to electrostatic or steric interactions. Van der Waals forces may also become important and lead to thinning when the surfaces approach each other quite closely. Hydrodynamic instability analyses have been used to estimate the rupture times for films under these conditions. Much work continues in this area.

As the foam rises, drainage occurs and some mass transfer may occur. Bubbles formed within the bulk liquid phase rise, collapse, and coalesce due to gas diffusion and the forces mentioned above. The enrichment of the foam phase relative to the exiting liquid phase depends on a number of factors including the adsorption of the surface-active solute, the bubble size, the liquid and gas flow rates, the initial concentration of the surface-active species, and the pH and ionic strength.

A number of experimental configurations have been discussed for foam separation (4). These include batch and continuous operations (see Fig. 1). In a batch operation, a solution of the surface-active species is stripped of the species by introduction of bubbles from a sparger. The rising foam is collected and coalesced to produce the overhead stream. Some of the coalesced

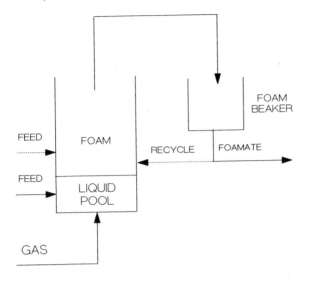

CONTINUOUS MODE

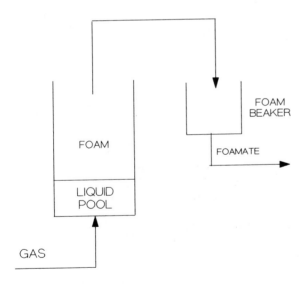

BATCH MODE

FIG. 1  Experimental configurations for foam separation.

foam may be returned to the foam column similar to that in a
batch distillation column.  In a continuous operation, the entering liquid may be introduced into the liquid pool or into the
foam section of the column.  Overhead product is drawn off at
the top of the column and underflow from the liquid pool.
Some of the overhead may be recycled analogous to a distillation column.  In the majority of laboratory studies, simple
batch operation or continuous introduction of liquid into the
liquid pool is performed.

## II.  THEORETICAL DEVELOPMENT

In the early papers on foam separation, an ideal foam model
was employed.  This model assumes the following (11,12):
foam is composed of spherical bubbles, no bubble breakage in
the foam (i.e., complete foam stability), the liquid entrained is
of the same composition as that of the bulk liquid, the liquid
pool is well mixed, equilibrium exists between the bulk liquid
and the interface concentrations, and the surfactant concentration is below the critical micelle concentration.  Based on the
last assumption, the surface concentration may be determined
from the Gibbs adsorption equation:

$$d\sigma = -RT\Sigma_i \; \Gamma_i \; d \ln a_i \tag{1}$$

where $\sigma$ is the interfacial tension, R is the gas constant, T the
absolute temperature, $\Gamma_i$ is the surface excess of component i,
$a_i$ is the activity of component i.  For low concentrations where
the activity equals the concentration, the surface excess is approximately constant and can be determined from a plot of the
surface tension as a function of the logarithm of the concentration.  (See Fig. 2, which is a plot of surface tension as a
function of the logarithm of the concentration of the surfactant
sodium dodecyl sulfate.  The surface excess is approximately
constant for surfactant concentrations between 20 and 150 ppm,
by weight.)  The reader is directed to any of the excellent
texts on interfacial phenomena for a more detailed discussion of
the Gibbs equation (13,14) as well as recent papers (15,16).
     The surface excess $\Gamma$ divided by the bulk phase concentra-
tion $C_i$ is a quantity called the distribution coefficient that is a
measure of the extent of separation.  A plot of the distribution
coefficient as a function of concentration shows a region where

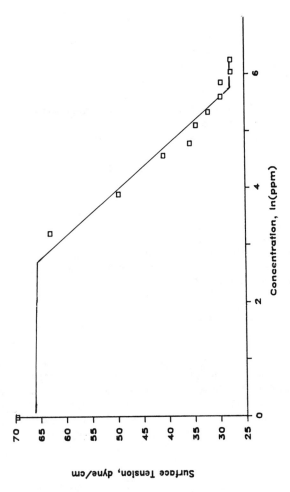

FIG. 2   Surface tension vs. concentration:   sodium dodecyl sulfate.

the distribution coefficient goes through a maximum or remains constant for low concentrations and abruptly decreases toward zero for higher concentrations (7, 17). Figure 3 is a plot of the surface excess calculated from the data on Fig. 2, divided by the bulk concentration of sodium dodecyl sulfate surfactant, vs, the surfactant concentration. The abrupt decrease at high concentrations is a consequence of the formation of micelles in the bulk phase as the concentration increases. Further increases in concentration of the surfactant end up with the formation of more micelles while the surface excess remains constant. Above the critical micelle concentration (CMC) separation drops radically.

Material balances for an ideal continuous foam separation column can be made to relate the liquid feed rate F and concentration $C_f$; the flow rate of collapsed foam, O the overhead product, with concentration $C_o$; the flow leaving the liquid pool, B the bottoms product, with a concentration $C_b$:

$$F = O + B \tag{2}$$

$$FC_f = OC_o + BC_b \tag{3}$$

The rate of solute leaving the column with the foam can be related to that carried in the liquid in the lamella and that adsorbed on the surface, i.e.,

$$OC_o = OC_b + SA\Gamma \tag{4}$$

where S is the volumetric foam generation rate, A is the gas-liquid interfacial area per unit volume of foam, and $\Gamma$ is the surface excess or surface concentration. Note that no further mass transfer of solute to the interface during foam rise is assumed in this model and the surface excess or concentration is constant. If spherical bubbles are assumed, A can be related to bubble diameter D and the gas flow rate G by

$$A = 6G/(DS) \tag{5}$$

These expressions can be combined to yield the following:

$$C_o = C_b + G(6/D)\Gamma/O \tag{6}$$

$$C_o/C_b = 1 + G(6/D)\Gamma/(C_bO) \tag{7}$$

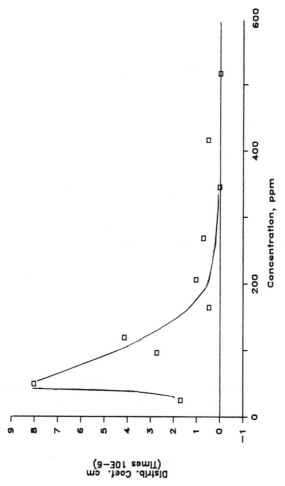

FIG. 3    Distribution coefficient vs. concentration:    sodium dodecyl sulfate.

$$OC_o/(FC_f) = [OC_b + G(6/D)\Gamma]/FC_f \tag{8}$$

These expressions are for the foamate concentration $C_o$, the enrichment ratio $R = C_o/C_b$, and the separation fractional efficiency $E = OC_o/FC_f$. The expressions can be further manipulated to result in

$$C_o = C_f + (6/D)\Gamma(G/O)(1 - O/F) \tag{9}$$

$$R = C_o/C_b = \frac{C_f(F/B) + (6/D)\Gamma(G/O)}{C_f(F/B) - (6/D)\Gamma(G/B)} \tag{10}$$

$$E = OC_o/(FC_f) = O/F + (\Gamma/C_f)(6/D)(G/B)(B/F)^2 \tag{11}$$

These expressions indicate the effect of a variety of parameters on the separation. For instance, above a certain concentration, the surface concentration $\Gamma$ becomes constant, so for an increase in feed concentration above this the foam concentration will approach the feed concentration and the enrichment ratio will decrease toward unity. This is shown in experiments with the enzyme glucoamylase being foam-separated from its fermentation broth. Figure 4 (18) depicts the enrichment ratio for the separation of the enzyme as a function of the enzyme activity or concentration in the liquid pool in a batch column run. The separation efficiency will decrease to the ratio of the foamate flow to the feed flow.

A decrease in bubble size provides more surface area for adsorption and consequently an increase in the concentration in the foam. The enrichment ratio and separation efficiency will also rise. An increase in the gas flow rate will result in an increase in the foamate concentration because of the increase in surface area per volume of processed liquid that results. However, for high gas flow rates, more liquid is entrained so that the foam liquid flow rate increases as well. These effects oppose each other so an optimum gas flow rate may be possible. A similar conclusion is reached concerning the enrichment ratio. Whereas it would appear that an increase in the gas flow rate, by material balance considerations, would increase the separation efficiency, actual experimental results often indicate a decrease due to a wetter foam. Figure 5 (18) depicts the enrichment ratio for foam separation of glucoamylase from its fermentation broth as a function of the gas flow rate.

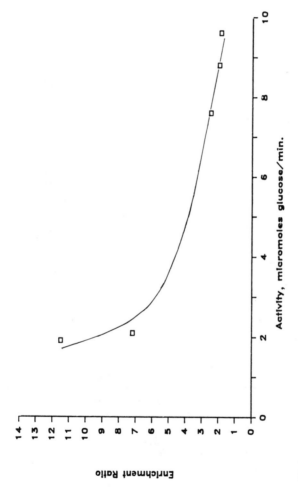

FIG. 4   Foam separation of glucoamylase:   enrichment ratio vs.  activity.

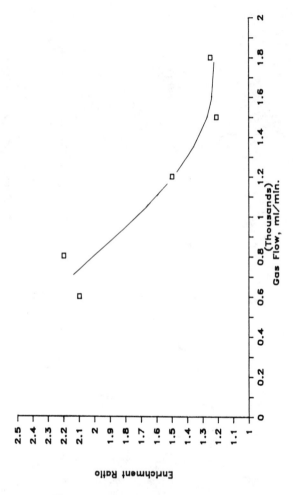

FIG. 5    Foam separation of glucoamylase:    enrichment ratio vs.   gas  flow.

The foam flow rate depends on a number of things including the gas flow rate, column dimensions, bulk, and surface properties of the liquid. Much research has gone into the determination of foam stability and estimation of the foam density (amount of liquid passing up through the column with the exiting gas). An increase in the inlet liquid flow rate would result in a decrease in the foamate concentration from a material balance consideration assuming that the foam flow rate did not change. The enrichment ratio and separation efficiency would also decrease.

A wetter foam, i.e., an increase in the foamate flow rate with no change in gas flow rate, results in a lower foamate concentration provided the foamate flow rate is much lower than the liquid flow leaving the pool (generally the case). A decrease in enrichment ratio is also predicted, whereas an increase in the separation efficiency due to material balance considerations is predicted. These conclusions are qualitatively correct for most operating systems as seen above and noted below.

The experimental results of Wood and Tran (11) indicate only qualitative agreement between the experiment and theory. They also found that increasing the column diameter increases both the foam drainage rate and foamate concentration. Note that the ideal foam model does not include the effects of column diameter or liquid pool and foam height.

The use of the ideal foam model (12) has been extended by Lemlich (4) and Jashnani and Lemlich (19) to allow calculation of the number of theoretical stages for various column operating conditions and a surface excess that is constant, varies linearly with concentration, or has the form of a Langmuir isotherm. The authors present extensive experimental information that allows the determination of the height of a transfer unit. Experimental data of their own and others indicate that the foam column represents several equilibrium stages whereas the liquid pool can be represented as one equilibrium stage. The height of a transfer unit varied from around 10 cm to greater than 1000 cm, increasing as the viscosity of the bulk phase increased.

For a detailed determination of the rates of surfactant transport and adsorption, the reader is directed to a recent paper by Borwankar and Wasan (20) in which bulk liquid diffusion and adsorption is modeled for a surfactant that follows a Frumkin isotherm.

## III.  EXPERIMENTAL STUDIES AND APPLICATIONS

### A.  Foam Stability

Foam stability depends on a number of factors, such as dynamic
surface tension, surface viscosity, surface elasticity, the electrical double layer, the viscosity of the bulk phases, gas diffusion, and solubility (8,21,22).  A discussion of the equation
of state of a foam was presented by Morrison and Ross (23).
This equation relates the volume, pressure, temperature, surface
tension, surface area, and moles of gas, and is the basis for
most theoretical expressions of foam stability.

Foam emerging from a liquid pool is fairly wet and is called
*kugelschaum* (24).  Interstitial liquid rapidly drains from these
foams by gravity until the liquid lamella become thin and surface tension forces begin to dominate.  Drainage then occurs
due to the lower pressure in the liquid at the junction of three
gas bubbles.  This region is called a plateau border.  Liquid
flows from the lamella formed between two gas bubbles into the
plateau border from which it then drains by gravity.  As the
film drains and the lamella thickness decreases, other factors
become important for dry foams.

Gas diffusion from smaller bubbles into larger due to the
capillary pressure difference between the bubbles of different
radii results in the growth of larger bubbles at the expense of
the smaller.  In addition, as the walls of the lamella approach
each other, electrical repulsion from the electrical double layers
becomes important and results in the "disjoining pressure" of
Derjaguin which can stabilize foams.

In some cases sudden rupture and collapse of bubbles occur.
Gibbs elasticity, the production of a surface stress due to surface tension gradients which oppose bubble shrinkage, is
thought to promote bubble stability.  Recent results (25), however, indicate that monolayer collapse is a decisive factor in
determining bubble stability.  Experimental evidence indicates
that as a monolayer is compressed beyond its collapse pressure,
bubble lifetimes decrease dramatically.  Reservoirs of collapsed
monolayer capable of rapid respreading are thought to result in
the decrease in bubble stability.

Good discussions of experimental techniques for determining
surface viscosities and elasticities have been presented by
Kanner and Glass (10) and Shah et al. (26).  These authors
also discuss the effect of surface viscosity on foam stability.
Surface viscosities may be enhanced by the presence of mixed

surfactants, especially alcohols and alkylsulfates. The enhanced surface viscosities increase the drag on liquid leaving the lamella and result in increased foam stability. The improvement in foam stability at the lower concentrations of surfactant may also lead to improved separation of a given surfactant when combined with another of an opposite charge (27). As these authors note, however, the effect may be reversed at high concentrations.

Bubbles in dry foams can also acquire polyhedral shapes that result in their being termed "polyederschaum" foams (24). These foams, due to their relatively flat surfaces, lose liquid mainly by capillary pressure rather than by gas diffusion. A recent note on the stability of plane parallel films has been published by Manev et al. (28).

Mathematical expressions for foam drainage and thinning can be found in texts by Davies and Rideal (13), Bikerman (29), and Rosen (30). More current models concentrate on gas diffusion (31) and gas diffusion with thinning by capillary flow (32–34). Numerical solution for the equations of the latter model with one adjustable constant appear to correlate with the experimental data (32). More complicated expressions including several equations, film properties, and column dimensions and operating conditions have been derived to allow prediction of foam flow rate (36–39) for dry foams.

A considerable number of studies have been conducted by Wilson and others (9) on the electrical properties of foams. These studies indicate that the Guoy-Chapman theory may be used to predict a streaming potential formed along the length of the foam column. Theoretical predictions and experiments indicate that the streaming potential varies directly with the column length, the $\zeta$ potential, and inversely with electrolyte concentration and conductivity. This has many ramifications in the area of foam flotation of colloids (40,41).

As a sidelight, a number of papers discuss the use of electrical conductivity to measure bubble size and velocities in both bubbles dispersed in a continuous liquid (42) as well as the foam density for a liquid dispersed in a gas (19,43–45). In the former papers, bubble diameters of 1 mm with velocities 0.05–0.3 m/sec have been measured by this technique. In the latter papers, bubbles measured photographically of sizes down to 0.08 mm are reported (46) for the operation of a 5.1 cm-diameter column, 1 m tall with aqueous solutions of anionic, cationic, and nonionic surfactants. The bubble size affects the foam density (fraction of foam volume occupied by liquid), which

is directly related to the conductivity. In this study, a uni-
modal distribution of bubble sizes was noted for most of the
runs except with albumin where a bimodal distribution resulted
from the presence of satellite bubbles. The foam ascended the
column in plug flow. Coalescence was indicated by a increase
in the spread of bubble size distribution for some of the systems. For most of the systems with surfactant concentrations
above the critical micelle concentration, moderate coalescence
was observed to occur by gas diffusion from the smaller
bubbles to the larger rather than by rupture. Consequently,
the foam density did not vary too much.

## B.  Biological Materials

This class of surface-active materials includes proteins that
have large numbers of hydrophobic and hydrophilic groups.
These molecules are quite large and diffuse to the surface
only slowly relative to smaller surface-active molecules such as
oleic acid, sodium dodecylbenzene sulfonate, etc. Another dis-
tinguishing feature of foam separation of proteins such as
enzymes is the possible denaturation of the enzyme due to the
change in configuration it takes at the surface relative to the
bulk phase. Foams produced by agitation generally contain
denatured enzymes (47). Nevertheless, as indicated in the
cited reference, foams have been used to separate a variety
of enzymes over the last 50 years. Emphasis has been on
foam formation near the isoelectric point of the enzyme, use of
nonionic surfactants to promote foam formation, and, if denaturation is possible, low gas flows and gentle conditions.
Many enzymes were found to be stable under foaming conditions,
however, such as catalase and amylase (48) and urease (49).
    Foams produced by proteins such as albumin are often
quite stable and dry, which promotes separation of the protein
from the bulk solution (50). A discussion of the separation of
albumin from its solution and the parameters affecting the
separation is presented by Ahmad (51). Separation was found
to be best at a pH near the isoelectric point (pH 4.9). The
foam stability was highest near this pH. The enrichment ratio
was seen to decrease from greater than 2 to less than 1 with
increasing albumin concentration from 50 to 500 ppm. The
enrichment ratio was found to decrease with increasing gas rate
presumably due to a wetter foam produced. The enrichment
was also seen to decrease with an increase in the liquid flow
rate. The enrichment increased with an increase in foam

height perhaps due to a drier foam at the higher height as well as more surface area available for adsorption. The effect of these and other parameters such as temperature (increasing enrichment with temperature), bubble size, pool depth, and other additives upon the separation of albumin is discussed by Gehle and Schugerl (52,53) and Schnepf and Gaden (54).

Other biological substances have been separated by foams. These include penicillin (52), viruses (55), and bacteria (56-66). The interaction between proteins, especially membrane proteins, and surfactants such as hexa(oxyethylene) dodecyl ether without loss of activity has resulted in the use of surfactants to extract proteins from membranes or cells. A brief paper by Nishikido et al. (67), presents isotherms for the complex formation. It has been shown that there is a maximum in the number of surfactant molecules bound per protein molecule as a function in surfactant concentration. This point is near the critical micelle concentration of the surfactant (68).

## C. Surfactant Removal

In many cases foams are produced in operating plants and hinder the operation. For instance, in wastewater treatment plants, man-made and natural surfactants produce foams in the aerated lagoons. These foams are often unsightly and may hinder operation. Distillation and gas absorption towers sometimes experience similar problems which lead to short circuiting in the towers.

The high surface concentrations of surfactants in the foam have resulted in the use of foams to separate synthetic detergents from wastewaters (4,69). Alkyl benzene sulfonate has been effectively removed from a series of municipal and industrial wastewaters in work reported by Brunner and Stephan (69). Removals of 50-70% were obtained in foam columns. A 500,000 gal/day pilot plant study is also reported in which a trough-type foam separator was able to achieve 70% removal of alkylbenzenesulfonate from an entering stream containing around 1-10 ppm. Some removal of other organic species was observed. A 10 million gal/day plant was estimated to cost 1.7 cents per daily gallon (capital costs) without including disposal of the foam and 1.9 cents/1000 gal operating cost (69).

Spent sulfite liquor from a paper-pulping process was foam-fractionated to remove 90% of the surface-active species such as fatty acids, resins, and lignosulfonates (70) as well as 40% of the toxics (resin acids and phenolics). [Seventy-one other papers concerned with foam separation of pulping products include those by Brasch and Robilliard (72) on kraft black liquor and by Keirstead and Caverhill (73) on calcium ligno-sulfonate.]

A novel aspect of foam separation of surfactants is the use of foams to separate surfactants and particulates from seawater blowdown brines (74). Neodol 25-3A (Shell Chemical Company) is an anionic surfactant that is added to seawater to improve its wettability and consequently the heat transfer coefficient in boiler tubes. About 95–97% of the surfactant was removed as well as some particulate material in a 5000 gal/day foam separation column.

The use of foams to remove typical surfactants is discussed by Newson (12) and Kubota et al. (75). The influence of molecular weight of nonionic surfactants on their removal by foaming is presented in a paper by Zwierzykowski et al. (76).

## D.  Ion Separation

Heavy metal ions such as zinc and cadmium cations or chromate or plumbate anions may be separated by the formation of a complex with a surface-active species. The surface-active species should generally have a charge opposite to that of the metal ion. For instance, 4-dodecyl diethylenetriamine can exchange a hydrogen atom to form a complex in the ratio of two molecules of surfactant to one of metal with the metal ions cadmium, nickel, or copper (+2) cations (77). This complex will exhibit surface activity. If two metallic species are being complexed and separated in the foam, both the complexation equilibrium and the surface adsorption must be considered. Chou and Okamoto (77) and Karger and Miller (17) present the following relationship for the relative distribution coefficient $\alpha$ for the two species, a and b, complexing with the surfactant s:

$$\alpha_{ab} = \frac{(\Gamma/C)_{as}[C_{as}/(C_a + C_{as})]}{(\Gamma/C)_{bs}[C_{bs}/(C_b + C_{bs})]} \tag{12}$$

where $C_a + C_{as}$ is the total amount of species a present. If the reversible reaction a + s = as is assumed, $K_a = C_{as}/(C_s C_a)$, and the above expression can be written as

$$\alpha_{ab} = \frac{(\Gamma /C_{as})[K_a/(K_aC_s + 1)]}{(\Gamma /C_{bs})[K_b/(K_aC_s + 1)]} \tag{13}$$

The chelation constants K were found to decrease from $10^{13}$ to $10^8$ in the order copper, nickel, cadmium. Chou and Okamoto (77) noted that the presence of surfactant in excess of that required by metal ion chelation resulted in nickel being selectively removed over copper due to the higher surface excess of nickel. When there is insufficient surfactant, copper is removed preferentially due to the higher stability constant.

The separation of other cations is discussed in papers by Jacobelli-Turi et al. (78) (uranium and thorium by benzethonium chloride); Siy and Talbot (79) (zinc by sodium dodecyl benzene sulfonate); Kubota et al. (80—82) (cadmium by sodium dodecyl benzene sulfonate); Jurkiewicz (83) (cadmium by hexdecyltrimethylammonium chloride); Okamoto and Chou (84) (cadmium and copper by 4-dodecyldiethylenetriamine); Huang et al. (85) (chromium by reaction with metal hydroxide precipitates and flotation with sodium lauryl sulfate surfactant); Kubota et al. (86) (magnesium and cadmium with sodium dodecyl benzene sulfonate); Kubota and Hayashi (87) (sodium, cadmium, and chromium ions with sodium dodecyl benzene sulfonate). An interesting study of the complex formed between copper and cobalt and octanoic and hexanoic acids is presented by Dodd and Jones (88). These authors evaluate the chemical and physical structure of the complexes. A good review of cation separation by foams is that of Valdes-Kreig et al. (89). The authors use the ideal foam separation model to explain trends in the experimental data and discuss the effect of pH, electroselectivity, mass transfer, column design, and scale-up.

Metal oxyanions such as rhenium (VII), molybdenum (VI), chromium (VI), tungsten (VI), and vanadium (V) may also be removed by complexation of the oxyanion with cationic surfactant such as hexadecyldimethylbenzylammonium chloride (90). These results indicate an optimum ratio of surfactant to metal ion concentration near the stoichiometric amount required for complexation. For concentrations above this, the uncomplexed surfactant competed with the complex for adsorption and below this there was insufficient surfactant to complex all of the metal. In some cases a precipitate is formed between the oxyanion and the surfactant and ion flotation occurs (91,92). Other anions include the cyanide complexes of silver and gold with hexadecyltriethylammonium iodide (93), chloride complexes of zinc, cadmium, mercury, and gold with hexadecyltrimethylammonium chloride (94), cyanide complexes of zinc, cadmium, mercury, and gold with hexadecyltrimethylammonium chloride (95), cyanide complexes of zinc, cadmium, and mercury with hexadecyltrimethylammonium iodide (96), cobalt with a variety of surfactants (97), chromate and thiosulfate ions with hexadecyldimethylammonium bromide (91).

An overview of ion separation is presented by Grieves (98). The subject is also discussed by Somasundaran (7,99) in his monographs. In some cases, as illustrated in these monographs, pH effects may include precipitate formation of metal ions and affect the distribution coefficient between metal ions.

## E.  Other Applications and Experimental Studies

An extensive list of examples of foam separation appears in the monograph by Somasundaran (7) and the review by Grieves (90). The reader is directed to these references as well as those discussed above for detailed discussions.

Concerning operating observations, Datye and Lemlich (45) found that the bubble distribution is not uniform across the foam column diameter. Smaller bubbles tend to concentrate at the column wall. This fact should be considered when photographs through the column walls are used to determine bubble sizes. In addition to the effects discussed above, vertical misalignment of the column can affect the results (100). A misalignment of 1 degree resulted in an increase in foamate concentration of 10–15% presumably due to back-mixing and perhaps better drainage for the shorter column lengths.

Most of the research papers on foams concern only one foaming agent. In a recent paper by Sharma et al. (101), the effect of chain length compatibility between two surfactants (sodium dodecyl sulfonate and a variety of alcohols, octyl, decyl, and docecyl) is presented. Where the chain lengths are compatible, the surface tension goes through a minimum, the surface viscosity a maximum, the bubble size a minimum. In addition, a foam produced from such a mixture has optimal fluid displacement properties (of interest in tertiary oil recovery from oil fields).

Shah and Mahalingam (102) discuss the use of a foam column as a gas absorption reactor. The authors considered the case of a very fast reaction [carbon dixoide (10% in air) absorption into a foam consisting of water, a nonionic or cationic surfactant,

and sodium hydroxide] and a slow reaction (carbon dioxide
absorption by sodium carbonate-bicarbonate solution with surfactant). The authors conclude that the increase in surface
area of up to 8 times that for packed or plate towers does not
overcome the presence of a interfacial resistance to mass transfer (perhaps due to reduced hydrodynamic flow) in the foam
column. For the high reaction rate case, the mass transfer
rates for the foam column are substantially below those for a
packed column. For the slow reaction case, however, the
rates in the foam column are 2—4 times higher than in a conventional absorption tower.

## F.  Bubble Fractionation and Solvent Sublation

In the case where a stable foam is not formed but surface-
active species are present, bubble fractionation may be possible (4,103). In this case a tall narrow column is used to
reduce back-mixing and air is injected at the bottom. As the
bubbles rise, surfactant is adsorbed on their surface. When
the bubbles collapse at the top of the liquid pool, a region of
high surfactant concentration results. A mathematical model
of a batch system with back-mixing has been developed by
Lemlich (1).

Solvent sublation, nonfoaming adsorptive bubble separation,
originated with Sebba (8,104,105) and employs another liquid
phase instead of or in addition to air upon which the solute
adsorbs. Minute gas bubbles surrounded by a soapy film,
called colloidal gas aphrons by Sebba (105), have been used to
remove metal ions. Solvent sublation has been used to separate
a number of organic dyes (106) and chemicals as well as metal
ions. An ideal stagewise calculation model is presented by
Clarke and Wilson (8) as well as aspects of mass transfer.


## IV.  FUTURE DEVELOPMENTS

Future theoretical and laboratory scale work will probably consist of the incorporation of the theoretical and experimental
results on foam stability and characteristics with the current
rather simple models for adsorptive separation. In addition,
new mass transfer expressions may also be developed to extend
the current models.

Some recommendations for future work are described by
Clarke and Wilson (8). These include pilot work in the more

promising foam or solvent sublation separation processes. They also recommend the use of baffles or column design to reduce back-mixing or axial dispersion in the liquid pool and thus obtain better separation and use of surfactant in the liquid pool section of a column. The use of micro gas dispersion techniques pioneered by Sebba (104,105) also shows promise. These minute bubbles (in the order of 10 μm) or colloidal gas aphrons (105) have extremely large amounts of surface area and long residence times in liquid pools. The result is almost complete separation of the adsorbed solute. Similar results have been obtained with minute liquid droplets, so-called colloidal liquid aphrons (105). Other developments in the area of flotation of precipitates and ores are also discussed. The use of foams will continue to be pursued for the removal or separation of trace amounts of materials, primarily from water. Practical applications are limited since the foam stream containing the concentrated materials must subsequently be handled to remove these materials. Foam separation, then, will primarily be a means of concentration of materials prior to further processing. As such it will probably be limited to specialty chemicals such as biological chemicals, trace toxics such as heavy metals, and organics that are not economically concentrated or separated by other means.

## REFERENCES

1. R. Lemlich, *AICHE J.*, *12*: 802 (1966).
2. B. L. Karger, R. B. Grieves, and R. Lemlich, *Sep. Sci.*, 2: 401 (1967).
3. B. L. Karger and D. G. DeVivo, *Sep. Sci.*, 3: 393 (1968).
4. R. Lemlich (ed.), *Adsorptive Bubble Separation Techniques*, Academic Press, New York, 1972.
5. R. Lemlich, *Ind. Eng. Chem.*, 60: 16 (1968).
6. T. A. Pinfold, *Sep. Sci.*, 5: 379 (1970).
7. P. Somasundaran, in *Separation and Purification Methods*, Vol. 1 (E. S. Perry and C. J. van Oss, eds.), Marcel Dekker, New York, 1972, p. 117.
8. A. N. Clarke and D. J. Wilson, *Foam Flotation: Theory and Applications*, Marcel Dekker, New York, 1983.
9. A. N. Clarke and D. J. Wilson, *Sep. Sci.*, *10*: 371 (1975).
10. B. Kanner and J. E. Glass, *Ind. Eng. Chem.*, *61*: 31 (1969).

11. R. K. Wood and T. Tran, *Can. J. Chem. Eng.*, *44*: 322 (1966).
12. I. H. Newson, *J. Appl. Chem.*, *16*: 43 (1966).
13. J. T. Davies and E. K. Rideal, *Interfacial Phenomena*, Academic Press, New York, 1963.
14. K. L. Mittal and E. J. Fendler (eds.), *Solution Behavior of Surfactants*, Vols. I and II, Plenum Press, New York, 1982.
15. V. T. Zharov, A. I. Rusanov, and S. A. Levichev, *Colloid Surf.*, *2*: 37 (1981).
16. R. Perea-Carpio, F. Gonzalez-Cabellero, J. M. Bruque, and C. F. Gonzalez-Fernandez, *J. Colloid Interf. Sci.*, *110*: 96 (1986).
17. B. L. Karger and M. W. Miller, *Anal. Chim. Acta, 48*: 273 (1969).
18. T. E. Carleson, "Foam Separation of Cells and Enzymes from Fermentation Broths," Presented at Annual AICHE Meeting, San Francisco, Calif., November 1984.
19. I. L. Jashnani and R. Lemlich, *Ind. Eng. Chem. Process. Des. Dev.*, *12*: 312 (1973).
20. R. B. Borwankar and D. T. Wasan, *Chem. Eng. Sci.*, *38*: 1637 (1983).
21. J. J. Bikerman, *Ind. Eng. Chem.*, *57*: 56 (1965).
22. S. Ross, *Chem. Eng. Prog.*, *63*: 41 (1967).
23. I. D. Morrison and S. Ross, *J. Colloid Interf. Sci.*, *95*: 97 (1983).
24. S. Ross, *Ind. Eng. Chem.*, *61*: 48 (1969).
25. R. L. Ternes and J. C. Berg, *J. Colloid Interf. Sci.*, *98*: 471 (1984).
26. D. O. Shah, N. F. Djabbarah, and D. T. Wasan, *Colloid Polym. Sci.*, *256*: 1002 (1978).
27. S. N. Hsu and J. R. Maa, *Ind. Eng. Chem. Process Des. Dev.*, *24*: 38 (1985).
28. E. D. Manev, S. V. Sazdanov, and D. T. Wasan, *J. Colloid Interf. Sci.*, *97.* 595 (1984).
29. J. J. Bikerman, *Foams*, Springer-Verlag, New York, 1973.
30. M. J. Rosen, *Surfactants and Interfacila Phenomena*, Wiley, New York, 1978.
31. H. C. Cheng and R. Lemlich, *Ind. Eng. Chem. Fundam.*, *24*: 44 (1985).
32. G. M. Nishioka, S. Ross, and M. Whitworth, *J. Colloid Interf. Sci.*, *95*: 435 (1983).
33. G. Nishioka and S. Ross, *J. Colloid Interf. Sci.*, *81*: 1 (1981).

34.  A. Monsalve and R. S. Schechter, *J. Colloid Interf. Sci.*, 97: 327 (1984).
35.  R. A. Leonard and R. Lemlich, *AICHE J.*, 11: 25 (1965).
36.  R. A. Leonard and R. Lemlich, *AICHE J.*, 11: 18 (1965).
37.  F. S. Shih and R. Lemlich, *AICHE J.*, 13: 751 (1967).
38.  P. A. Haas and H. F. Johnson, *Ind. Eng. Chem. Fund.*, 6: 225 (1967).
39.  D. Desai and R. Kumar, *Chem. Eng. Sci.*, 38: 1525 (1983).
40.  D. J. Wilson, *Sep. Sci.*, 12: 231 (1977).
41.  D. J. Wilson, *Sep. Sci.*, 12: 447 (1977).
42.  D. A. Lewis, R. S. Nicol, and J. W. Thompson, *Chem. Eng. Res. Des.*, 62. 334 (1984).
43.  K. S. Chang and R. Lemlich, *J. Colloid Interf. Sci.*, 73: 224 (1980).
44.  A. F. Sharovarnikov, V. A. Kokushkin, and E. V. Kokorev, *Kolloidn. Zh.*, 42: 406 (1980).
45.  A. K. Datye and R. Lemlich, *Int. J. Multiphase Flow*, 9: 627 (1983).
46.  I. L. Jashnani and R. Lemlich, *Ind. Eng. Chem. Fund.*, 14: 131 (1975).
47.  A. Thomas and M. A. Winkle, in *Topics in Enzyme and Fermentation Biotechnology* (A. Wiseman, ed.), Ellis Horwood Ltd., Chichester, Sussex, England, 1977.
48.  S. E. Charm, J. Morningstar, C. C. Matteo, and B. Paltiel, *Anal. Biochem.*, 15: 498 (1966).
49.  M. London, M. Cohen, and P. B. Hudson, *Biochem. Biophys. Acta*, 13: 19 (1954).
50.  Z. Lalchev and D. Exerowa, *Biotechnol. Bioeng.*, 23: 669 (1981).
51.  S. I. Ahmad, *Sep. Sci.*, 10: 673 (1975).
52.  R. D. Gehle and K. Schugerl, *Appl. Microbiol. Biotechnol.*, 19: 373 (1984).
53.  R. D. Gehle and K. Schugerl, *Appl. Microbiol. Biotechnol.*, 20: 133 (1984).
54.  R. W. Schenpf and E. I. Gaden Jr., *J. Biochem. Microbiol. Technol. Eng.*, 1: 1 (1959).
55.  M. D. Guy, J. D. McIver, and M. J. Lewis, *Water Res.*, 10: 737 (1976).
56.  P. R. Fields, P. J. Fryer, N. K. H. Slater, and G. P. Woods, *Chem. Eng. J.*, 27: B3 (1983).
57.  H. W. Bretz, S. L. Wang, and R. B. Grieves, *Appl. Microbiol.*, 14: 778 (1966).
58.  A. M. Gaudin, A. L. Mular, and R. F. O'Connor, *Appl. Microbiol.*, 8: 91 (1960).

59. A. J. Rubin, F. A. Cassel, O. Henderson, J. D. Johnson, and J. C. Lamb III, *Biotechnol. Bioeng.*, *8*: 135 (1966).
60. A. J. Rubin, *Biotechnol. Bioeng.*, *10*: 89 (1968).
61. A. M. Gaudin, N. S. Davis, and S. E. Bangs, *Biotechnol. Bioeng.*, *4*: 211 (1962).
62. A. M. Gaudin, A. L. Mular, and R. F. O'Connor, *Appl. Microbiol.*, *8*: 84 (1960).
63. A. M. Grieves, N. S. Davis, and S. E. Bangs, *Biotechnol. Bioeng.*, *4*: 223 (1962).
64. R. B. Grieves and S. L. Wang, *Biotechnol. Bioeng.*, *8*: 323 (1966).
65. R. B. Grieves and S. L. Wang, *Appl. Microbiol.*, *15*: 76 (1967).
66. R. B. Grieves and S. L. Wang, *Biotechnol. Bioeng.*, *9*: 187 (1967).
67. N. Nishikido, T. Takahara, H. Kobayashi, and M. Tanaka, *Bull. Chem. Soc. Jpn.*, *55*: 3085 (1982).
68. M. N. Jones, P. Manley, and P. J. W. Midgley, *J. Colloid Interf. Sci.*, *82*: 257 (1981).
69. C. A. Brunner and D. G. Stephan, *Ind. Eng. Chem.*, *57*: 40 (1965).
70. J. E. Jajic, D. Berk, and L. A. Behie, *Can. J. Chem. Eng.*, *57*: 321 (1979).
71. D. Berk, J. E. Zajic, and L. A. Behie, *Can. J. Chem. Eng.*, *57*: 327 (1979).
72. D. J. Brasch and K. R. Robilliard, *Sep. Sci. Technol.*, *14*: 699 (1979).
73. K. F. Keirstead, L. Caverhill, and D. DeKee, *Can. J. Chem. Eng.*, *60*: 680 (1982).
74. E. Valdes-Krieg, C. J. King, and H. H. Sephton, *Desalination*, *16*: 39 (1975).
75. K. Kubota, A. Yasui, and S. Hayashi, *Can. J. Chem. Eng.*, *55*: 96 (1977).
76. W. Zwierzykowski, K. B. Medrzycka, and S. Chlebus, *Sep. Sci.*, *10*: 381 (1975).
77. E. J. Chou and Y. Okamoto, *Sep. Sci. Technol.*, *13*: 439 (1978).
78. C. Jacobelli-Turi, A. Barocas, and S. Terenzi, *Ind. Eng. Chem. Process Des. Dev.*, *6*: 161 (1967).
79. R. D. Siy and F. D. Talbot, *Can. J. Chem. Eng.*, *55*: 67 (1977).
80. K. Kubota, S. Hayashi, and M. Morita, *Can. J. Chem. Eng.*, *56*: 130 (1978).

81. K. Kubota, S. Hayashi, and Y. Takubo, *Can. J. Chem. Eng.*, *57*: 591 (1979).
82. K. Kubota, *Can. J. Chem. Eng.*, *53*: 706 (1975).
83. K. Jurkiewicz, *Sep. Sci. Technol.*, *20*: 1979 (1985).
84. Y. Okamoto and E. J. Chou, *Sep. Sci.*, *11*: 79 (1976).
85. S. D. Huang, C. F. Fann, and H. S. Hsieh, *J. Colloid Interf. Sci.*, *89*: 504 (1982).
86. K. Kubota, H. Kawanoue, and S. Hayashi, *Can. J. Chem. Eng.*, *55*: 101 (1977).
87. K. Kubota and S. Hayashi, *Can. J. Chem. Eng.*, *55*: 286 (1977).
88. J. W. Dodd, A. D. Jones, and J. Parkinson, *J. Inorg. Nucl. Chem.*, *40*: 1797 (1978).
89. E. Valdes-Kreig, C. J. King, and H. H. Sephton, *Sep. Purif. Methods*, *6*: 221 (1977).
90. R. B. Grieves, *Chem. Eng. J.*, *9*: 93 (1975).
91. R. B. Grieves, R. L. Drahushuk, W. Walkowiak, and D. Bhattacharyya, *Sep. Sci.*, *11*: 241 (1976).
92. R. B. Grieves, D. Bhattacharyya, and J. K. Ghosal, *Colloid Polym. Sci.*, *254*: 507 (1976).
93. W. Walkowiak and Z. Rudnik, *Sep. Sci. Technol.*, *13*: 127 (1978).
94. W. Walkowiak, D. Bhattacharyya, and R. B. Grieves, *Anal. Chem.*, *48*: 975 (1976).
95. W. Walkowiak and R. B. Grieves, *J. Inorg. Nucl. Chem.*, *38*: 1351 (1976).
96. T. Gendolla and W. A. Charewicz, *Sep. Sci. Technol.*, *14*: 659 (1979).
97. K. Shakir and M. Aziz, *Chem. Scr.*, *11*: 164 (1977).
98. R. B. Grieves, *AICHE Symp. Ser.*, *71*: 143 (1975).
99. P. Somasundaran, *Sep. Sci.*, *10*: 93 (1975).
100. E. Valdes-Kreig, C. J. King, and H. H. Sephton, *AICHE J.*, *21*: 400 (1975).
101. M. K. Sharma, D. O. Shah, and W. E. Brigham, *Ind. Eng. Chem. Fundam.*, *23*: 213 (1984).
102. P. S. Shah and R. Mahalingam, *AICHE J.*, *30*: 924 (1984).
103. K. Kubota, *Can. J. Chem. Eng.*, *53*: 706 (1975).
104. F. Sebba, *Ion Flotation*, Elsevier, Amsterdam, 1962.
105. F. Sebba, *Sep. Purif. Methods*, *14*: 127 (1985).
106. G. N. Shah and R. Lemlich, *Ind. Eng. Fund.*, *9*: 350 (1970).

# 11

# Mineral Separation by Froth Flotation

DOUGLAS  W.  FUERSTENAU and RONALDO HERRERA-URBINA*
Department of Materials Science and Mineral Engineering,
University of California at Berkeley, Berkeley, California

---

*Present affiliation: Saltillo Institute of Technology, Saltillo, Coahuila, Mexico.

## SYNOPSIS

Froth flotation is the separation process whereby small mineral
particles are captured by air bubbles and removed from the
slurry as a froth. Consequently, two broad types of reagents
are necessary in flotation: reagents to control the wettability
of mineral particles and reagents to control foaming or frothing.
Effecting selective mineral separations is accomplished by the
selective adsorption of surfactants at the solid-liquid interface.
The emphasis in this chapter is the delineation of the surface
chemical phenomena responsible for the adsorption of surfactants at the mineral-water interface. How adsorption affects
wettability, particle-bubble interaction, and flotation response
is discussed in detail for various kinds of surfactants and
mineral systems. How various other kinds of reagents assist
in the control of wetting behavior and flotation response is also
presented.

## I.  INTRODUCTION

Froth flotation is a surfactant-based separation process that
has become the single most important operation used today in
the mineral industry throughout the world.  In the United
States, for example, more than one-third of the mineral concentrates produced are obtained by froth flotation (1).  In
general, the froth flotation process is applied to separate both
metallic and nonmetallic minerals, and to upgrade solid fuel
materials.  A recent mineral industry survey conducted by the
United States Bureau of Mines indicated that in 1985 domestic
mineral-processing plants which employed froth flotation
processed more than 422 million short tons of ore and raw
coal to produce over 80 million tons of mineral concentrates (2).
This required almost 1.4 billion pounds of chemical reagents,
with surfactants accounting for 33% of this total or about
$459 \times 10^6$ lb.  Surfactants used as flotation reagents include
those that function as collectors, depressants, frothers,
flocculants, dispersants, and filtering aids.

   The froth flotation process separates individual mineral or
solid particulates from a mixture of finely divided solids suspended in aqueous solutions; it involves attachment of certain
solid species to gas bubbles introduced into the system and
the subsequent removal of the froth containing the desired
solids.  The successful separation of solid particles by froth
flotation depends primarily on differences in the affinity for
water molecules exhibited by their surfaces.  Hydrophobic
materials possess surfaces that are not wetted by water and
they readily attach to gas bubbles, while hydrophilic solids
strongly interact with water and thus have no affinity for the
gas phase.  With the exception of coal, which has a carbona-
ceous skeleton, and such materials as graphite, sulfur, talc,
pyrophyllite, molybdenite, and stibnite, commercial minerals
have hydrophilic surfaces that must be rendered hydrophobic
if their separation by froth flotation is to be achieved.

   Although many processes enter into play in froth flotation,
those that most influence its performance involve the modification of the wetting characteristics of mineral surfaces and the
attachment of solid particles to gas bubbles.  The wetting
behavior of the surface of minerals dispersed in aqueous solutions is generally modified upon the addition of collector-acting
surfactants into the suspension.  Having a strong tendency to
adsorb at the solid-liquid interface, collector ions or molecules

are able to displace adsorbed water molecules from the interfacial region, thus rendering the solid surface hydrophobic.
Collectors of interest in flotation can be classified into two
types: (1) long-chain ionic surfactants that adsorb physically
as counterions in the electrical double layer and (2) surfactants
that chemisorb or chemically react at the interface. The first
class of collectors is characterized by a strong hydrocarbon
chain—hydrocarbon chain interaction between adsorbed ions in
the Stern layer.

In addition to collectors, the effective flotation separation
of minerals from their ores requires the use of depressants
that inhibit the flotation of undesired mineral particles. The
beneficiation of iron ores by froth flotation using different
methods will be used to illustrate how the flotation separation
of minerals is accomplished. In the case of iron ores, the
usual problem involves separating hematite from quartz. Table
1 lists six different procedures that have been reported to
achieve this separation (3). The first four procedures have
been used industrially. Separation of quartz from hematite by
flotation using amines at pH 6—7 is possible because this collector adsorbs only on the negatively charged quartz particles,
but not on the essentially uncharged hematite particles. This
separation is also possible commercially by using starch to
depress hematite, and calcium ions to activate quartz, which is
then floated at pH 11—12 with sodium oleate. In the third
process, hematite responds to flotation at pH 2—4 upon adsorption of the anionic alkyl sulfonate collector on its positively
charged surface. Under these conditions, quartz is negatively
charged and the anionic surfactant does not adsorb on its surface. The fourth industrial process involves chemisorption of
oleate on hematite but not on quartz. Chemisorption of a
chelating agent, namely octyl hydroxamate, on hematite also
allows its flotation separation from quartz because this reagent
does not interact with silica. In this procedure methyl isobutyl
carbinol is used as a frother. Finally, hematite activated with
either hydrochloric acid or sulfuric acid can be separated from
quartz if an amine is used as the flotation collector at pH 1.5.

Once the desired mineral has been made selectively hydrophobic through the adsorption of collectors, gas bubbles are
introduced into the suspension to effect its separation from the
hydrophilic species, which remain suspended in the air-water-
mineral dispersion. The froth flotation process, therefore,
takes place under highly turbulent conditions whereby both
gas bubbles and mineral particles are suspended in a stirred

TABLE 1    Use of Surfactants to Separate Hematite ($Fe_2O_3$)
and Quartz ($SiO_2$) by Froth Flotation

| Surfactant added and solution conditions | Mineral floated |
| --- | --- |
| Alkyl amine salt, pH 6—7 | Quartz |
| Sodium oleate and starch, pH 11—12 in the presence of calcium ions | Quartz |
| Sodium alkyl sulfonate, pH 2—4 | Hematite |
| Sodium oleate, pH 6—8 | Hematite |
| Potassium octyl hydroxamate and methyl isobutyl carbinol, pH 8.5 | Hematite |
| Alkyl amine salt, pH 1.5 in the presence of hydrochloric acid or sulfuric acid | Hematite |

aqueous suspension.  The factor that determines the success of
this operation is the interaction between gas bubbles and mineral particles under such a dynamic state of the system.  Of particular importance are the forces acting on a rising gas bubble
and their effect on bubble-particle collision and subsequent
attachment of the solid to the gas phase.  Particle-bubble
attachment under the hydrodynamic conditions present in froth
flotation is facilitated by the addition of frother-acting surfactants into the system (4).  Three elementary steps are involved in this dynamic attachment:  (1) bubble-particle collision
with the formation of a thin film of liquid between them; (2)
thinning of the film to the point of rupture; and (3) rupture
and recendence of the film, giving rise to a stable bond
between the gas bubble and the solid particle.
    A complete understanding of froth flotation must incorporate
both the physical and chemical characteristics of the solid-water,
solid-gas, and water-gas interfaces.  Several thermodynamic
aspects relating to interfacial phenomena at the mineral-aqueous
solution and mineral-aqueous solution-gas bubble boundaries
will be presented in this chapter.  The kinetics of gas bubble—
mineral particle interaction will also be discussed briefly.  The
principles to achieve selective separation of minerals by froth
flotation will be analyzed in terms of the surface chemical and
physical phenomena of a variety of pure mineral-collector
systems.  These will include oxides, silicates, sparingly soluble

minerals, and soluble salts in the presence of both cationic
and anionic collectors.

## II. ELECTRICAL ASPECTS OF THE MINERAL/AQUEOUS SOLUTION INTERFACE

Since nearly all flotation reagents are added to the aqueous
mineral suspension before the gas bubbles are introduced into
the pulp, it is indispensable to have as much information as
possible about the physicochemical characteristics of the
mineral-aqueous solution interface in order to gain a better
understanding of the mechanisms of surfactant adsorption at
that interface and its effect on mineral flotability. One of the
aspects of this interface that most influences the flotation
behavior of minerals is its electrical nature. The interfacial
electrical properties not only control collector adsorption on
minerals but also the state of aggregation or dispersion of
mineral suspensions, and the stability of thin liquid films
between gas bubbles and mineral particles.

### A. The Electrical Double Layer

Mineral particles dispersed in aqueous suspensions carry a
distinct electric charge on their surfaces. To maintain the
electrical neutrality of the system, an excess charge develops
in the solution region close to the surface. This excess solution charge, which is equal in magnitude but opposite in sign
to the surface charge, develops in a diffuse region extending
out into the solution. In the simplest case, the surface charge
and the diffuse region of solution charge constitute the electrical double layer at the interface.

### B. Electrical Charge and Potential on Mineral Surfaces

The mechanisms responsible for the development of the electric
surface charge at the mineral-aqueous solution interface may
involve not only the dissociative adsorption of water molecules
but also preferential dissolution of lattice ions, specific adsorption of surface active ions, and isomorphous substitution of ions
comprising the mineral lattice. The surface charge density $\sigma_0$
is defined in terms of the adsorption density of both potential-
determining cations $\Gamma_+$ and anions $\Gamma_-$ as

$$\sigma_0 \equiv zF(\Gamma_+ - \Gamma_-) \tag{1}$$

where F is the Faraday constant and z is the valence for a
symmetrical (1−1 or 2−2) salt. Potential-determining ions are
defined as those ions that have the ability to control the magnitude of both the surface charge and the surface electrical
potential $\psi_0$, which is the electrical potential difference between
the mineral surface and the bulk solution. The activity of the
potential-determining ions in solution at which the surface
charge density is zero gives the point of zero charge (PZC).
Table 2 lists the pH values for the PZC of a number of oxide,
silicate, and ionic minerals commonly separated by froth flotation. These values provide the most important information for
describing the surface behavior of minerals in aqueous
suspensions.

The surface potential at any activity of potential-determining ions is given by

$$\psi_0 = \frac{RT}{zF} \ln \frac{a_+}{(a_+)_{PZC}} \tag{2}$$

where R is the gas constant, T is the absolute temperature,
and $a_+$ is the activity of the potential-determining cation. At
the PZC, and in the absence of specific adsorption, the surface electrical potential is zero. Since absolute potentials
across the interface cannot be measured, the electrical potential of the bulk aqueous phase is generally assigned a value
of zero (10).

As a consequence of the development of an electric charge
on mineral surfaces upon their immersion in aqueous solutions,
certain ions present in the bulk solution must concentrate at
the solid-liquid interface to balance the surface charge and
maintain electroneutrality in the system. These ionic species
are called counterions and they build up an electrical double
layer, consisting of a Stern layer and a diffuse layer that
spreads out into the solution. The distribution of counterions
in this double-layer region is controlled by electrical attractive forces and thermal motion.

The closest distance of approach of a counterion to the
surface is determined by the size of hydrated ions that adsorb
at the surface forming a compact layer. As a result, the first
layer of counterions lies at a distance $\delta$ from the surface.
This boundary between the compact layer of ions adsorbed at
the surface and the diffuse layer of counterions is termed the
Stern plane, and its potential is taken to be $\psi_\delta$. Figure 1

**TABLE 2**   Point of Zero Charge of Some Oxide, Silicate, and Ionic Minerals

| Mineral | pzc (pH) | Ref. |
|---|---|---|
| $SiO_2$, α-quartz | 2-3 | 5 |
| $Fe_2O_3$, hematite (natural) | 4.8-6.7 | 5 |
| $Fe_2O_3$, hematite (synthetic) | 8.6 | 5 |
| $Al_2O_3$, corundum | 9.1 | 5 |
| $FeOOH$, goethite | 6.8 | 5 |
| $MnO_2$, pyrolusite | 5.6 | 6 |
| $BaSO_4$, barite | 3.4 | 7 |
| $CaF_2$, fluorite | 6.2 | 7 |
| $CaCO_3$, calcite | 9.5 | 8 |
| $MnSiO_3$, rhodonite | 2.8 | 6 |
| $CuSiO_3 \cdot 2H_2O$, chrysocolla | 2.0 | 6 |
| $LiAl(SiO_3)_2$, spodumene | 2.6 | 9 |

illustrates schematically the electrical double layer showing the surface charge, the location of the Stern plane, and the diffuse layer of counterions.  This figure also shows that in the absence of specific adsorption, the potential drop from the surface to the Stern plane is linear, and from this plane on out into the solution it falls off gradually to zero in the bulk solution.  The electrical potential at any point from the interface varies considerably depending on such parameters as the ionic strength, the nature of the counterions, and the concentration in solution of potential-determining ions.  Counterion adsorption at the interface may arise from simple electrical attraction of a combination of electrostatic and specific adsorption mechanisms.  If the only driving force for the adsorption of the ions of a particular compound is electrical in nature, this chemical reagent is called an indifferent electrolyte.  Some ions, however, exhibit surface activity in addition to electrostatic interaction, and they can adsorb through such mechanisms as covalent bond formation, hydrogen bonding, hydrophobic adsorption, and solvation effects.

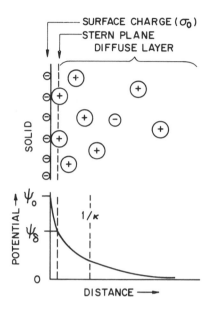

FIG. 1    Schematic representation of the electrical double
layer and the potential drop across the double layer.

## C.   The Diffuse Electrical Layer and the $\zeta$ Potential

Electrical double-layer phenomena have been widely investigated
in flotation research by measurements of the well-known electrokinetic or $\zeta$ potential.   Electrokinetic effects are the result
of relative motion between the solid and liquid phases in a
charged system.   This relative movement of the fluid with
respect to the solid causes transport of counterions in the diffuse layer outside the slipping plane.   Even though the exact
position of the shear plane has not been unequivocally located,
detailed studies with micelles indicate that the slipping plane
occurs very close to the surface, so that taking $\zeta$ as a measure
of $\psi_\delta$ is a reasonably good approximation (11).   The location of
the slipping plane at the interface and the variation of potential through the double layer under various conditions is shown
in Fig. 2.   Double-layer potentials can be controlled either by
changing the concentration of potential-determining ions or by
finding conditions where inorganic or organic ions specifically
adsorb at the interface.

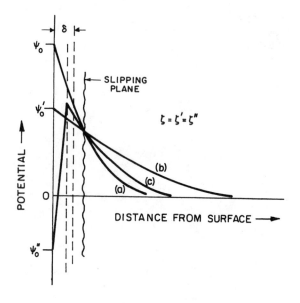

FIG. 2    The variation of potential through the double layer
(a) for a high concentration of potential-determining ion, (b)
for a low concentration of potential-determining ion, and (c) in
the presence of a specifically adsorbed counterion with the surface charge reversed.


    Perhaps the most commonly used technique to measure the
$\zeta$ potential at the mineral-aqueous solution interface is electrophoresis.   This electrokinetic technique involves measurement
of the movement of dispersed particles when the suspension is
placed in an electrical field.   Electrophoretic mobility u can
then be converted into $\zeta$ potential using the following expression:

$$u = \frac{\varepsilon \zeta}{4\pi\eta} \tag{3}$$

where $\eta$ and $\varepsilon$ are the viscosity and the dielectric constant of
the liquid, respectively, and $\pi = 3.1416$.   To accurately determine $\zeta$ potentials from electrophoretic results, and particularly
for very small particles, one must use a functional relation
between particle radius and the inverse of the thickness of the

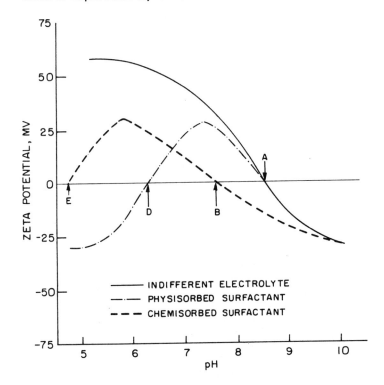

FIG. 3    Schematic representation of the effect of surfactant adsorption on minerals on their $\zeta$ potential at the solid-liquid interface.

diffuse layer of counterions such as that given by Wiersema et al. (12).

Under certain conditions, surface-active counterions may be able to reverse the sign of the $\zeta$ potential. The solution condition at which $\zeta$ reverses sign has been termed the isoelectric point (IEP) or simply the point of $\zeta$ potential reversal (PZR) (13). The $\zeta$ potential changes brought upon the adsorption of surfactants at the interface are shown schematically in Fig. 3 where point A corresponds to the PZC of the material and points B, D, and E are points of $\zeta$ reversal but not points of zero charge. Fuerstenau (14) described in detail the close relationship between adsorption phenomena and the electrical double layer at mineral-aqueous solution interfaces.

## III.  WETTING OF MINERALS BY AQUEOUS SOLUTIONS

The wetting characteristics of mineral surfaces play a predominant role in controlling their flotation behavior.  After the
collision of a particle with a gas bubble, they will determine
whether or not attachment occurs.  Wetting phenomena refer
to the spontaneous displacement of a fluid from a solid surface
by a second immiscible fluid (15), and they are generally classified as adhesional wetting, spreading wetting, and immersional
wetting.  These three wetting processes can be described by
thermodynamic relations for the change in surface free energy
per unit area $\Delta G_i$ for the immersion of a dry solid particle into
a liquid medium is given by

$$\Delta G_i = \gamma_{SL} - \gamma_{SG} \tag{4}$$

where $\gamma_{SL}$ and $\gamma_{SG}$ are the surface tensions at the solid-liquid
and solid-gas interfaces, respectively.  Spontaneous wetting of
the solid by the liquid will take place only if $\Delta G_i$ is negative,
i.e., if the surface tension of the solid-liquid interface is
lower than that of the solid-gas interface.  Flotation, there-
fore, is favored by relatively low $\gamma_{SG}$ and high $\gamma_{SL}$.  Low $\gamma_{SG}$
values are characteristic of hydrophobic solids whose surfaces
have a nonpolar character, exhibit low energy, and are not
wetted by water.  In general, minerals having at least one
crystal plane held together by van der Waals forces only exhibit natural flotability.  Important examples are graphite,
molybdenite, pyrophyllite, stibnite, and talc.  Most minerals,
however, possess high-energy surfaces that strongly interact
with water molecules, causing them to be completely wetted by
water.  In this section, the thermodynamics of wettability,
flotability, and the contact angle that forms when the solid,
liquid, and gas phases meet will be discussed.

### A.  Wettability and Flotability

The wettability of minerals by water is commonly assessed by
comparing the work of adhesion of water to their surfaces $W_a$
with the work of cohesion of water $W_c$.  The work of adhesion
is defined as the work per unit of surface required to create
a solid-gas and liquid-gas interface upon destroying a solid-
liquid interface:

$$W_a = \gamma_{SG} + \gamma_{LG} - \gamma_{SL} \tag{5}$$

where $\gamma_{SL}$ is the surface tension at the solid-liquid interface. The work of cohesion of water is given by

$$W_c = 2\gamma_{LG} \qquad (6)$$

In terms of these quantities, froth flotation is favored if the work of adhesion of water to the mineral is smaller than the work of cohesion of the water molecules. Fowkes (16) suggested that the work of adhesion is composed of several additive terms, the most important being the contribution due to van der Waals dispersion forces $W_a^d$, hydrogen bonding forces $W_a^h$, and electrical forces $W_a^e$:

$$W_a = W_a^d + W_a^h + W_a^e \qquad (7)$$

Since dispersion forces are always present at interfaces, $W_a^d$ is usually the dominant term in Eq. (7). Moreover, an ideal hydrophobic solid (such as paraffin wax) is defined as a material for which the work of adhesion of water involves only dispersion forces, and Fowkes's treatment shows that

$$W_a = W_a^d \cong 2\sqrt{\gamma_L^d \gamma_S^d} \qquad (8)$$

where $\gamma_L^d$ and $\gamma_S^d$ are the dispersion force contributions to the surface energies of the liquid and the solid, respectively. The $\gamma^d$ values reported by Fowkes (17) for several materials are tabulated in Table 3 along with some values calculated by Laskowski and Kitchener (18) of the dispersion force contribution to the work of adhesion of water to these materials. Since for none of these materials is $W_a^d$ as large as the work of cohesion of water, Laskowski and Kitchener (18) concluded that all solids would be hydrophobic if they did not carry polar groups.

Froth flotation can also be described in terms of the change in free energy of the system. Before bubbles make contact with mineral particles in an aqueous minpral suspension, the surface free energy of the system, on a unit area basis, is given by

$$G_0 = \gamma_{SL} + \gamma_{LG} \qquad (9)$$

TABLE 3    Dispersion Force Contribution to the Surface
Energy and the Work of Adhesion of Water for Several
Representative Materials

| Substance | $\gamma^d$ (erg $\cdot$ cm$^2$) | $W_a^d$ (erg $\cdot$ cm$^2$) |
|---|---|---|
| Water | 21.8 | — |
| Paraffin wax | 25.5 | 98 |
| Iron oxide | 107 | 97 |
| Graphite | 110 | 98 |
| Silica | 123 | 104 |
| Rutile | 140 | 110 |
| Mercury | 200 | 132 |

The free energy change accompanying the displacement of
water molecules from the solid-water interface by a gas phase
upon contact of the gas bubbles with mineral particles in the
flotation pulp is expressed as

$$\Delta G_f = \gamma_{SG} - (\gamma_{SL} + \gamma_{LG}) \tag{10}$$

The attachment of a gas bubble to a mineral particle suspended
in an aqueous solution gives rise to the formation of a three-
phase boundary, whose most distinct characteristic is the contact angle.

## B.  Thermodynamics of the Contact Angle

Assessment of mineral wettability has traditionally been accomplished by contact angle measurements.  The application of
this technique to physicochemical research in froth flotation
was brought to a high level by the extensive efforts of Wark
and his coworkers (19).  The contact angle $\theta$ is defined as the
angle formed across the liquid phase when a gas bubble is
attached to the surface of a solid submerged in an aqueous
solution (Fig. 4).  It is measured across the liquid at the
three-phase contact line, and its magnitude is given by the
Young-Dupre equation, which defines the equilibrium contact
angle in terms of the three interfacial tensions:

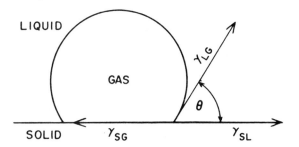

FIG. 4 Schematic representation of the contact angle on a solid immersed in an aqueous solution.

$$\cos \theta = \frac{\gamma_{SG} - \gamma_{SL}}{\gamma_{LG}} \tag{11}$$

Using this equation to modify the expression for the change in free energy of the immersional wetting process, namely, Eq. (4), we obtain

$$\Delta G_i = -\gamma_{LG} \cos \theta \tag{12}$$

The change in free energy for the adhesional dewetting process, which controls the successful separation of solid particles by froth flotation, has already been given by Eq. (10), but it can also be expressed as

$$\Delta G_f = \gamma_{LG} (\cos \theta - 1) \tag{13}$$

Thus, both immersional wetting and adhesional dewetting can be assessed in terms of two measurable quantities: the surface tension of the liquid and the equilibrium contact angle. Equation (13) shows that the change in free energy upon attachment of a gas bubble to a solid particle will be negative for any finite value of the contact angle. The general thermodynamic condition for flotation to occur would then be a finite contact angle. With the exception of a handful of minerals, however, clean mineral surfaces in contact with water exhibit zero contact angle, i.e., they are wetted by water. In order for these minerals to become flotable, their surfaces require a

modification of their wetting behavior. This is accomplished
by the adsorption of surfactants at the solid-liquid interface.

## C. Surfactant Adsorption and Dewetting Phenomena

Dewetting of mineral surfaces immersed in aqueous solutions is
accomplished upon the addition of certain surfactants into the
suspension. Surfactant ions that effectively adsorb at the
solid-liquid interface cover the reactive polar surface sites and
expose their hydrophobic part to the aqueous phase. When
surfactants that have been added to aqueous suspensions containing hydrophilic mineral particles adsorb at the interfaces,
they change the interfacial tensions and cause the contact angle
to become finite. At constant temperature T and pressure P,
the change in surface tension due to the adsorption of surfactant species x at any interface is given by

$$d\gamma = -RT\Gamma_x d (\ln a_x) \tag{14}$$

where R is the gas constant, $\Gamma_x$ is the surfactant adsorption
density in $mol/cm^2$, and $a_x$ is the activity of surfactant species
x in solution. In this expression, the adsorption density of
solvent water molecules is considered to be zero (Gibb's con-
vention), and the adsorption of surfactant species is referred
to this standard state. Wettability control and surface dewetting phenomena in flotation systems can then be interpreted in
terms of the adsorption density of surfactant species at interfaces and changes in interfacial tensions. Differentiating the
Young-Dupre equation with respect to ln $a_x$, at constant T and
P, and combining it with the expressions of surface tension
change due to surfactant adsorption at the solid-liquid, liquid-
gas, and solid-gas interfaces, the following relationship is
obtained (20):

$$\Gamma_x^{LG} \cos \theta = (\Gamma_x^{SG} - \Gamma_x^{SL}) + \left(\frac{\gamma_{LG}}{RT}\right) \left[\frac{d(\cos \theta)}{d(\ln a_x)}\right] \tag{15}$$

where the superscripts for the surfactant adsorption density
refer to the respective interfaces. This expression gives the
criterion for flotation in terms of collector adsorption. If the
contact angle increases with increasing collector activity (as is
the case for flotation systems involving relatively low surfactant
concentrations and minerals that are originally hydrophilic),

$d(\cos\theta)/d(\ln a_x)$ in Eq. (15) is negative, and the general
condition for stable bubble-particle attachment in flotation,
i.e., a finite contact angle, is accomplished only when
$\Gamma_x^{SG} > \Gamma_x^{SL}$. This general and important observation empha-
sizes that adsorption at all three interfaces present in froth
flotation systems must be considered in order to achieve a
better understanding of flotation phenomena (21–23). Moreover, since $\Gamma_x^{SG}$ must be greater than $\Gamma_x^{SL}$ for contact
angles to become finite, bubble-mineral attachment will depend
on the adsorption densities of all surfactant species present
in the system, namely, collectors and frothers. As discussed
by Fuerstenau and Raghavan (24), wettability and dewetting of
mineral surfaces is controlled primarily through the addition of
water-soluble surfactants to the aqueous solution followed by
their adsorption at the solid-liquid interface. Therefore, the
following paragraphs will discuss the mechanisms involved in
surfactant adsorption phenomena at this interface and analyze
the effect of various parameters on the adsorption process.

## IV.  SURFACTANT ADSORPTION ON MINERALS

Hydrophilic minerals can be separated from each other with
froth flotation by finding conditions for the selective adsorption
of organic surfactants, commonly known as collectors, onto
their surfaces. The property of surface activity of these reagents is due to the amphipathic or amphiphilic nature of their
molecules, which contain both a hydrophilic and a hydrophobic
part. The nature of the polar group on the molecule determines the type of its interaction with the mineral surface, while
the structure of the hydrocarbon chain controls the extent of
its interaction with water.

### A.  Thermodynamics of Surfactant Adsorption

A heterogeneous system is said to be in equilibrium when the
chemical potential of any species i is equal in all phases. In
the case of a collector species x present in a mineral-aqueous
collector suspension, its chemical potential in the bulk solution
$\mu_x$ is given by

$$\mu_x = \mu_x^{0} + RT \ln a_x \qquad (16)$$

where $\mu_x^0$ is its standard chemical potential. The chemical potential of the same species at the surface $\mu_x^s$ is

$$\mu_x^s = (\mu_x^0)^s + RT \ln a_x^s \tag{17}$$

where $(\mu_x^0)^s$ is its standard chemical potential at the surface and $a_x^s$ is its activity at the surface. Since $\mu_x = \mu_x^s$ at equilibrium,

$$\frac{a_x^s}{a_x} = \exp \left[ \frac{\mu_x^0 - (\mu_x^0)^s}{RT} \right] \tag{18}$$

Substituting the expression that defines the change in the standard free energy of adsorption $\Delta G_{ads}^0$, namely, $\Delta G_{ads}^0 = (\mu_x^0)^s - \mu_x^0$, in the foregoing equation and assuming that $a_x = C$ and $a_x^s = \Gamma_\delta /2r$, Eq. (18) can now be written in the form of the Stern-Grahame equation:

$$\Gamma_\delta = 2rC \exp \left( \frac{-\Delta G_{ads}^0}{RT} \right) \tag{19}$$

where C is the concentration of collector in solution, $\Gamma_\delta$ is its adsorption density in the Stern layer, and r is the effective radius of the adsorbed ion. This expression is strictly valid only at low collector concentrations because it does not include activity coefficients.

For rapid flotation to take place the collector ion must adsorb strongly in the Stern layer (25), and the collector adsorption requirements are no longer limited to low surface coverages. An expression for fractional adsorption densities at intermediate surface coverages can be derived by applying the law of mass action to a binary mixture of similarly sized molecules showing ideal behavior in both the liquid phase and in the adsorbed layer. The equilibrium constant K for this adsorption process is given by

$$K = \frac{x_1^s x_2}{x_2^s x_1} \tag{20}$$

where $X_1$ and $X_2$ refer to the mole fraction of solute and solvent, respectively, in the liquid phase, and $X_1^S$ and $X_2^S$ are their mole fractions in the adsorbed state. For a mineral-aqueous collector system, since $X_2^S = 1 - X_1^S$, Eq. (20) can be rewritten as

$$\frac{\alpha}{1 - \alpha} = \frac{C}{55.5} \exp\left(\frac{-\Delta G_{ads}^0}{RT}\right) \tag{21}$$

where $\alpha$ is the fraction of mineral surface sites covered by the collector species and C is the collector concentration in mol/liter.

The change in the standard free energy of adsorption of collectors can be expressed in terms of various contributions to the overall adsorption process:

$$\Delta G_{ads}^0 = \Delta G_{elec}^0 + \Delta G_{chem}^0 + \Delta G_{CH_2}^0 + \Delta G_{solv}^0 + \cdots \tag{22}$$

These individual terms represent the changes in the standard free energy due to electrostatic, chemical, hydrophobic bonding, and solvation effects, respectively. Depending on the mechanisms involved in the interaction of the collector with the mineral surfaces, the contributions to the change in the standard free energy of adsorption can be essentially zero or have a finite value.

## B.  Surfactant Adsorption Mechanisms

Collector species can adsorb at the mineral-aqueous solution interface as counterions in the Stern layer. In this case, the adsorption process involves only physical forces because the surfactant ion must be charged oppositely to the mineral surface in order to be attracted electrostatically. The forces responsible for specific adsorption are also physical and they are essentially those involved in the hydrophobic association between hydrocarbon chains:

$$\Delta G_{ads}^0 = \Delta G_{elec}^0 + \Delta G_{CH_2}^0 \tag{23}$$

When collector ions interact with surface sites on minerals to form new chemical compounds, the adsorption process is

chemical in nature and the change in the standard free energy
of the system is given by

$$\Delta G^0_{ads} = \Delta G^0_{elec} + \Delta G^0_{chem} \tag{24}$$

Depending on the magnitude of $\Delta G^0_{chem}$, the adsorption of
anionic collectors on minerals can occur at pH values considerably higher than their PZCs. For example, oleate can
chemisorb on hematite and hydroxamate on pyrolusite under
conditions where the surface charge is highly negative.

## C. Parameters Affecting Surfactant Adsorption

The nature of the surfactant molecule, the hydrocarbon chain
length, and the pH of the system are among the variables that
play a predominant role in collector adsorption phenomena at
mineral-aqueous solution interfaces.

### 1. Nature of Surfactant Molecule

The properties of both the polar group and the hydrophobic
chain of the surfactant molecule play an important role in controlling adsorption phenomena. The polar head of the molecule
will determine its ionization, hydrolysis, affinity for metal ions,
solubility of its salts, oxidation, solvation phenomena, and size
effects in electrostatic interactions. Surfactants used as flotation collectors are generally classified as cationic and anionic
if they yield positively charged or negatively charged surface-
active ions, respectively, upon dissociation in water. Cationic
surfactants will adsorb physically only on negatively charged
surfaces while anionics will adsorb similarly only on positively
charged surfaces. The hydrophobic chain of the molecule will
determine its solubility, critical micelle concentration, steric
effects in chemisorption, interaction of unsaturated bonds with
water molecules, and oxidizability of double bonds.

### 2. Effect of Surfactant Chain Length

Since the hydrocarbon chain length of surfactant ions controls
their interaction with water molecules, it has a pronounced
effect on surfactant adsorption at the mineral-aqueous solution
interface, thus affecting mineral flotation behavior. When the
concentration of a long-chain surfactant in solution is greater
than a critical value, ionic species tend to aggregate into

clusters called micelles.  Aggregation phenomena result in a decrease of hydrocarbon chain–water interactions, which are thermodynamically unfavorable.  Similar to the association of hydrocarbon chains in solution, surfactant ions adsorbed at the solid-liquid interface also begin to associate into two-dimensional aggregates after they reach a certain critical surface concentration.  This surface aggregation phenomenon was first proposed by Fuerstenau (26) and later by Gaudin and Fuerstenau (27), who also introduced the term "hemimicelle" to name these surface aggregates.

If the standard free energy for removing 1 mol of $CH_2$ groups from water through association is $\phi$, then the total contribution of hydrophobic bonding to the change in the standard free energy of adsorption is

$$\Delta G^0_{CH_2} = n\phi \qquad (25)$$

where $n$ is the number of $CH_2$ groups in the chain.  The effect of hydrocarbon chain length on collector adsorption can be indirectly delineated through $\zeta$ potential measurements, which indicate changes in the electrokinetic potential due to adsorption.  Using streaming potential techniques to study the dependence of the adsorption of alkylammonium ions at the quartz-aqueous solution interface on chain length, Fuerstenau (28) observed that at a certain concentration of alkylammonium ions the $\zeta$ potential of quartz undergoes a sudden change.  This phenomenon was attributed to the formation of hemimicelles.

Adsorption of organic ions in the Stern layer is related to their surface activity.  As the adsorption density in this layer increases, the distance between adsorbed ions decreases and they come sufficiently close for their hydrocarbon chains to associate.  The surfactant adsorption density at the onset of hemicelle formation is given by

$$\Gamma_\delta = 2rC \exp\left(\frac{-zF\psi_\delta - n\phi}{RT}\right) \qquad (26)$$

The magnitude of $\phi$ at the onset of hemimicelle formation has been evaluated from $\zeta$ potential measurements and Hallimond tube flotation results (3).  It was found to be about $-0.6$ kcal/mol of $CH_2$ groups, in agreement with values obtained from micelle formation studies and solubility data.

Mineral flotation with micelle-forming collectors should
exhibit a regular dependence on chain length, similar to the
well-known Traube's rule, because hemimicelle formation depends
on the length of the hydrocarbon chain of the collector. This
dependence of flotation response on chain length is clearly
seen in Fig. 5, where the results of quartz flotation with
amine collectors ranging from four to 18 carbon atoms are presented. It is interesting to note that complete flotation of
quartz requires about 1 M residual concentration of the four-
carbon aminium acetate while the required concentration to completely float quartz under the same conditions with the 18-carbon
aminium is only $10^{-5}$ M. The ability of long-chain surfactant
ions to function effectively as collectors at more dilute concentrations is due to their strong tendency to adsorb and aggregate through chain-chain interaction at the interface. Because
of the $n\phi$ term in Eq. (26), hemimicelle formation results in a
surface coverage greater than that due to electrostatic attraction alone.

## 3. Effect of pH

Surfactant adsorption at the mineral-aqueous solution interface
is affected by the pH because it affects both the electrical
nature of this interface and the distribution of surfactant
species in solution if the surface-active reagent is a weak acid
or base. In the case of slightly soluble minerals, the pH will
also control their solubility and this will in turn significantly
affect surfactant adsorption. Figure 6 presents adsorption isotherms that clearly show how the effect of pH on surface
charge influences the adsorption of two anionic surfactants,
namely, sodium dodecyl sulfonate and sodium oleate, on oxides.
In these systems, the pH does not affect the total concentration
of ionic surfactant species because these surfactants are completely dissociated in the pH range investigated. The adsorption of dodecyl sulfonate on positively charged alumina is a
physical process in which the negative sulfonate ions are attracted electrostatically to the alumina surface. As the pH is
raised, adsorption decreases and essentially ceases as the surface charge on alumina becomes negative above the PZC (pH
9.2). Oleate adsorption on hematite, on the other hand, involves a chemical mechanism that results in the formation of
ferric oleate at the surface. The adsorption isotherms in Fig. 6
indicate that oleate anions adsorb on hematite even when the
pH is two units above the PZC of hematite (pH 8.2). At

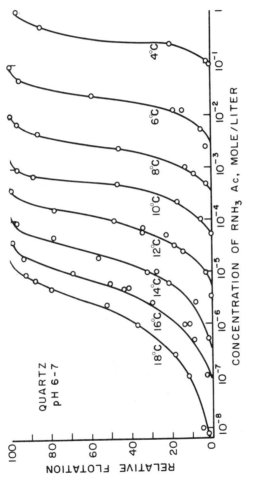

FIG. 5  Flotation response of quartz at pH 6–7 as a function of the concentration of alkylammonium acetates with various chain lengths. (From Ref. 29.)

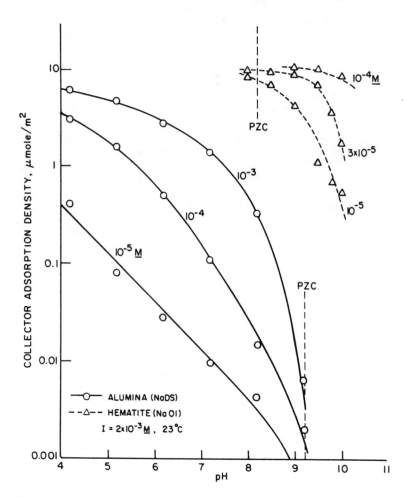

FIG. 6    Effect of pH on the physisorption of sodium dodecyl sulfonate on alumina and the chemisorption of sodium oleate on hematite.   (From Ref. 25.)

pH 10 and above, however, the surface is so negatively
charged that the $zF\psi_\delta$ term overcomes the strong effect of the
$\Delta G^0_{chem}$ term and oleate ions are repelled from the surface of
hematite.

In the case of electrolyte collectors that adsorb physically
on minerals, flotation phenomena are affected by the pH because
it also controls their solution chemistry. For example, the
flotation behavior of spodumene, which is negatively charged
above pH 2.6, with dodecylamine as the collector strongly
depends on the pH (30). In this system flotation occurs at
pH values above the PZC of the mineral and when the pro-
tonated amine predominates in solution, but it ceases upon complete hydrolysis of the collector. The positively charged
amine ion will predominate in solution below the $pK_a$ (the pH
value at which 50% of the electrolyte is ionized) of dodecylamine, which is around pH 10.6 (31). Above the $pK_a$, however, the concentration of cationic amine species decreases
considerably and flotation falls sharply, thus demonstrating
the negative effect of collector hydrolysis on mineral flotation
involving physical adsorption.

## D.  Correlation Between Collector Adsorption,
   $\zeta$ Potential, Contact Angle, and Flotation

Since the froth flotation process is a combination of various
phenomena occurring at three different interfaces, several
techniques must be used to quantify the effect of surfactant
adsorption on the properties of these interfaces and to relate
their changes to the flotation behavior of minerals. Figure 7
summarizes the results of collector adsorption density determinations, $\zeta$ potential and contact angle measurements, and
Hallimond tube flotation studies on the quartz-dodecylammonium
acetate (DAA) system (32). This figure clearly shows that
measurements reflecting conditions at the solid-liquid interface
(adsorption and $\zeta$ potential) can be correlated directly with
conditions at the more complex interfacial phenomena that occur
when solid, liquid, and gas phases meet (contact angle and
froth flotation). The most important feature from all of these
investigations is the sharp change in interfacial properties at a
certain collector concentration, which corresponds to the concentration necessary for the formation of hemimicelles. At
about $10^{-4}$ M DAA, there is an abrupt increase in the amount
of amine ions adsorbed at the interface. This collector

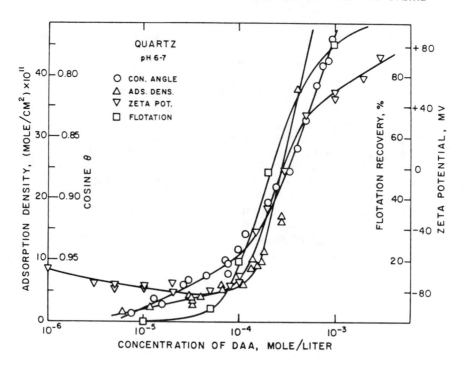

FIG. 7   Correlation of adsorption density, contact angle, and
$\zeta$ potential with the flotation of quartz at pH 6—7 as a function
of dodecylammonium acetate concentration.   (From Ref. 32.)

concentration corresponds to that at which the contact angle
increases significantly, and to the inflection point where the
$\zeta$ potential sharply becomes less negative.   Aminium ions are so
strongly adsorbed that they are able to reverse the sign of the
$\zeta$ potential upon a small increase in concentration.   From Fig. 7
we can also see that the onset of quartz flotation occurs under
conditions where hemimicelles begin to form, clearly indicating
that optimum flotation depends on strong adsorption in the
Stern layer.   These results are extremely useful because they
show that careful experimentation allows close correlation between collector adsorption density determinations, $\zeta$ potential
and contact angle measurements, and the flotation response of
minerals.

Certain mineral-collector systems, however, exhibit no correlation between the results of different experimental techniques
used in flotation research to evaluate changes in interfacial
properties due to surfactant adsorption. Several properties of
the mineral and the collector might be responsible for these
phenomena. For example, when minerals that release a significant amount of their lattice ions into solution come into contact
with aqueous solutions containing surfactants that interact
chemically with these ionic species, adsorption density determinations based in measurements of surfactant abstraction from solution will not reflect true adsorption at the interface. Precipitation of the collector in the bulk will result in high apparent
adsorption densities with insignificant or nil changes in contact
angle, $\zeta$ potential, and flotation behavior.

A typical example of a system exhibiting lack of correlation
between these four experimental techniques is the chrysocolla—
octyl hydroxamate system. The hydroxamate flotation response
of this copper silicate is low at pH values where both the contact angle and the collector uptake are high (33). In the case
of methylated quartz, Laskowski and Iskra (34) reported that
there is no correlation between contact angles and flotation
behavior. In relation to $\zeta$ potential changes with adsorption,
Laskowski and Kitchener (18) found that clean hydrophilic
silica exhibits practically the same $\zeta$ potential as that of methylated hydrophobic quartz.

## V.  GAS BUBBLE—MINERAL PARTICLE INTERACTION
IN FROTH FLOTATION

The concept of the contact angle, which has been extensively
used in flotation research to explain the wetting behavior and
flotation response of minerals, reflects the interaction of a
bubble and a solid surface under equilibrium conditions. This
static attachment, however, does not represent the dynamic
interaction between air bubbles and mineral particles that occurs
under highly turbulent conditions in flotation cells. Unfortunately, flotability studies based on dynamic attachment considerations present both experimental and theoretical difficulties
that have not been overcome. As a result, this area of flotation research is not fully understood at present. Both the
thermodynamic and kinetic aspects of bubble-particle interaction
will be discussed in this section.

## A.  Gas Bubble—Mineral Particle Collision

The first step leading to the attachment of particles onto
bubbles during froth flotation is the collision between these
two phases.  This process is controlled mainly by physical
parameters, of which the velocities of the bubble and the
particle are the most important.  Derjaguin and Dukhin (35)
appear to have been the first researchers to suggest that a gas
bubble moving upward in a liquid phase can be considered to
be surrounded by three distinct regions.  London—van der
Waals surface forces and electrical double-layer forces predominate at the liquid-gas interface.  Away from the surface,
diffusophoretic forces are important, while conventional fluid
mechanics controls the area outside the surface and the mass
transfer boundary layer.

As early as 1948, Sutherland (36) explained the collision
process in terms of hydrodynamic concepts that excluded the
inertial forces of the particle.  His hydrodynamic theory has
been confirmed by experimental investigations using strobo-
scopic methods to determine the efficiency of collision between
small particles and spherical air bubbles (37).  The collision
efficiency is a quantity that characterizes the ability of a bubble
to capture particles in a fluid flow field.  Upon collision, the
bubble undergoes an elastic deformation that is probably due
mainly to transformation of the kinetic energy of the particle.
The magnitude of the kinetic energy of the particle will
determine whether or not attachment occurs; high kinetic
energies will result in particle repulsion.  The time of contact
during collision (the collision time) is also critical in determining dynamic attachment, and it depends on the collision energy.
Thus, the combined action of the time and the energy of collision will determine the stability or rupture of the film of
liquid between a gas bubble and a mineral particle.

The effect of the presence of frother-acting surfactants at
the air-liquid interface of bubbles on collision times has been
recently reported by Leja and He (4).  They postulated that
the dipoles of frother molecules, owing to their ability to
polarize, can modify nearly instantaneously the interaction of
charges present at colliding interfaces during dynamic attachment.  The relaxation times of dipoles have to be shorter than
the collision times between particles and gas bubbles for their
attachment to be facilitated under dynamic conditions.

B.  Thinning of Liquid Films Between Solid and
    Gas Phases

The equilibrium thickness of thin films of aqueous solutions on
mineral surfaces plays a predominant role in froth flotation
because even after attachment there is always a layer of water
between the gas bubble and the mineral particle (38).  The
behavior of thin films on minerals is controlled by an excess
pressure, acting normal to the film, that opposes its thinning.

1.  The Disjoining Pressure

Thin-film phenomena have been described by Derjaguin and
Kusakov (39) in terms of the excess pressure, which they
called the disjoining pressure $\Pi$.  The disjoining pressure can
be expressed as the difference between the pressure within
the gas bubble p and the pressure in the bulk liquid $p_0$ adjacent to the film (40).

$$\Pi = p - p_0 \tag{27}$$

Wetting or nonwetting of a solid by a liquid will depend on
whether the disjoining pressure is positive or negative.  A
solid-liquid-gas system having a negative disjoining pressure
over a range of film thicknesses will exhibit finite contact
angles.  However, if the disjoining pressure of this system is
positive at all film thicknesses, the solid surface is covered by
a stable wetting film and the contact angle is zero.

   The thickness and stability of wetting films depend on various
components of the disjoining pressure.  Derjaguin et al. (41)
have given the following expression for the disjoining pressure:

$$\Pi = \Pi_e + \Pi_d + \Pi_a + \Pi_s \tag{28}$$

where $\Pi_e$, $\Pi_d$, $\Pi_a$, and $\Pi_s$ are the contributions of the ionic
electrostatic, molecular, adsorption, and structural components,
respectively.  The ionic electrostatic component results from
the overlapping of diffuse ionic layers, while the overlapping
of the zones of action of the van der Waals dispersion forces
gives rise to the molecular component.  The overlapping of
diffuse adsorption layers of surfactants, polymers, and neutral
molecules in solution controls the magnitude of the adsorption

component. Polar liquid molecules at solid-liquid interfaces
adopt a distinct structure, which is different from that of the
same type of molecules in the bulk. It is the deformation or
overlapping of these interfacial molecular layers that accounts
for the structural component of the total disjoining pressure.
In a more recent study of the kinetic theory of froth flotation,
however, Derjaguin and Dukhin (42) excluded the adsorption
component of the disjoining pressure, thus considering that
only three distinct forces contribute to the total disjoining
pressure.

In the case of an uncharged bubble and an ideal nonpolar
solid, Derjaguin and Dukhin (35) considered that bubble-
particle attachment is controlled only by electrostatic (repulsive) and van der Waals (attractive) forces. The electrostatic
force that hinders the thinning of the film between the bubble
and the particle is

$$\Pi_e = \frac{\varepsilon \psi_1^2}{8\pi\, d^2\, \cosh^2 (h/d)} \tag{29}$$

where $\varepsilon$ is the dielectric constant of the liquid, d is the double-
layer thickness, $\psi_1$ is the potential at the mineral surface, and
h is the film thickness. The van der Waals force for the
bubble-particle system is

$$\Pi_d = -\frac{A_{LL} - A_{LS}}{6\pi h^3} \tag{30}$$

where $A_{LL}$ and $A_{LS}$ are the Hamaker constants for molecules
of water interacting with each other and for water molecules
interacting with the solid, respectively. Using these relations,
Derjaguin and Shukakidse (43) calculated a flotation criterion
m that makes it possible to determine how the value of the $\zeta$
potential of a mineral particle controls its flotation behavior.
This flotation criterion is given by

$$m = \frac{d\, \varepsilon \zeta^2}{A} \tag{31}$$

where $A = A_{LL} - A_{LS}$. The relation of the flotability of
stibnite to its $\zeta$ potential is plotted in Fig. 8, which is based

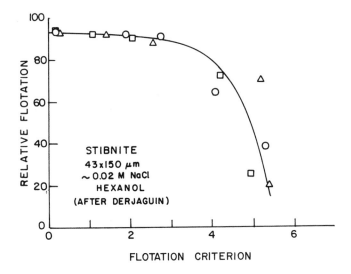

FLOTATION CRITERION

FIG. 8    Flotability of stibnite ($\Delta$) Kadan-Jai fields, ($\square$) Margusor fields, and ($\circ$) Zopkhito fields in 0.02 M NaCl in hexanol as a function of the flotation criterion m.

on data reported by Derjaguin and Shukakidse (43). These results indicate that optimum bubble-particle attachment in this system is achieved only when the flotation criterion is lower than 3, i.e., at relatively small $\zeta$ potentials.

2.   The Induction Time

The establishment of contact between an air bubble and a solid surface requires a definite period of time.   Upon collision the film of liquid between the particle and the bubble thins to a critical thickness.   The minimum time required for the film to drain to the critical thickness is defined as the induction time. Scheludko (44), taking into consideration van der Waals attractive forces, gives the following expression for the induction time $\tau_i$:

$$\tau_i \stackrel{\sim}{-} \frac{1}{h_{in}^2 \, aP} + \frac{12\pi}{aS} \, (h_{in} - h_c) \tag{32}$$

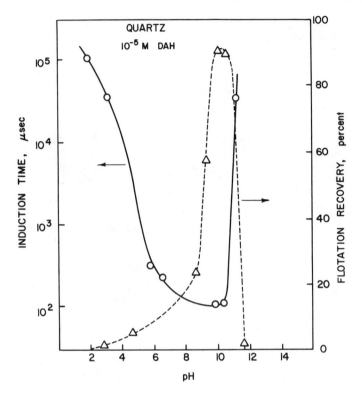

FIG. 9    Correlation of induction time with the flotation of
quartz as a function of pH in the presence of $10^{-5}$ M dodecylamine hydrochloride.

where a is the Laplace capillary constant, P is the capillary
pressure inside the gas bubble, S is the area of the liquid
film, and $h_{in}$ and $h_c$ are the initial and critical film thicknesses,
respectively.    The induction time is determined experimentally
under static conditions, and it is strongly affected by the type
and concentration of surfactant, ionic strength, temperature,
and particle size among other factors.    Lekki and Laskowski
(45) clearly established that frother additions effectively
decrease the induction time values.    Yordan and Yoon (46), on
the other hand, measured the induction time required for
bubble-particle attachment in the quartz—dodecylamine hydrochloride system under different conditions, and they correlated

these measurements with the amine flotation of quartz. Their
induction time measurements, presented in Fig. 9, show that
maximum flotation occurs under conditions of minimum induction times. This indicates that the wetting film must be
unstable for flotation to occur. It is now widely recognized
that the induction time of film rupture is of significant importance in evaluating the kinetics of froth flotation. Under
the dynamic conditions of flotation, the basic requirement for
gas bubble—mineral particle attachment is an induction time
smaller than the time of contact during their collision.

## VI.  FROTH FLOTATION OF MINERALS

There are several ways for achieving the separation of
minerals by froth flotation. The most common methods involve
flotation after direct physical or chemical adsorption of the
collector, flotation of activated minerals with collector ions
charged similarly to the surface, and flotation of mineral
fines either after flocculation or with coarse particles that act
as carriers.

### A.  Flotation by Physical Adsorption of
the Collector

The flotation behavior of minerals whose surfaces are made
hydrophobic upon physical adsorption of collectors is strongly
controlled by the PZC of the mineral and the solution pH.
Because physically adsorbing collectors function as counterions
in the electrical double layer at the mineral-water interface,
flotation takes place only if the active collector species is in
the ionic form and if the mineral surface and collector ion are
oppositely charged. Thus, minerals whose PZC values are
significantly different can be separated by froth flotation if
physisorbing collectors are used. Flotation by physisorption
of the collector, however, is not mineral-specific because
electrostatic interactions are the main driving force for adsorption. As a consequence, this type of adsorption does not provide the necessary conditions to achieve optimum mineral
selectivity in froth flotation of ores. Both anionic and cationic
collectors can adsorb physically on mineral surfaces.

### 1.  Mineral—Cationic Collector Systems

Chemical reagents most commonly used as cationic collectors for
the flotation of minerals are the salts of long-chain primary

amines. Since these surfactants ionize in aqueous solutions, the activity of the aqueous hydrogen ion controls the distribution of amine species in solution. In the case of dodecylamine, its solution chemistry is given by the following reactions (31):

$$RNH_3^+ = RNH_{2(aq)} + H^+ \qquad\qquad pK_a = 10.63 \qquad (33)$$

$$2RNH_3^+ = (RNH_3)_2^{2+} \qquad\qquad pK_d = -2.08 \qquad (34)$$

$$RNH_3^+ + RNH_{2(aq)} = RNH_3^+ \cdot RNH_2 \quad pK_{ad} = -3.12 \qquad (35)$$

$$RNH_{2(s)} = RNH_{2(aq)} \qquad\qquad pK_{sp} = 4.69 \qquad (36)$$

A speciation distribution diagram of the type presented as Fig. 10 can be constructed using these thermodynamic data. This diagram corresponds to a total concentration of $1 \times 10^{-4}$ M dodecylamine and shows that the amine is completely dissociated at acidic and neutral pH, but it hydrolyzes if the solution is made alkaline. It also gives the pH of precipitation and indicates the formation of dimer species and association between the charged amine ion and the neutral molecule. The associative interactions between ionized and neutral species, and their effect on precipitation and solution chemistry of amines, have been recently reported by Castro et al. (47). The $pK_a$ of the surfactant is an important parameter relevant to froth flotation involving physical adsorption processes. If the PZC of a certain mineral is known, identification of the pH regions where the cationic and neutral amine species predominate in solution will help in delineating the solution conditions necessary to achieve its separation from other minerals by froth flotation.

Both simple and complex oxide and silicate minerals, sparingly soluble minerals, and soluble salts have been found to respond to flotation with cationic collectors. It is well known that flotation of oxides using relatively low concentrations of amines usually reaches a maximum between pH 10 and 11. Based on the solution chemistry of dodecylamine, Ananthapadmanabhan et al. (31) correlate the cationic flotation to coincide with the pH of maximum concentration of iono-molecular dimers, namely, around pH 10. In a study of the flotation of quartz with $10^{-4}$ M dodecylamine, however,

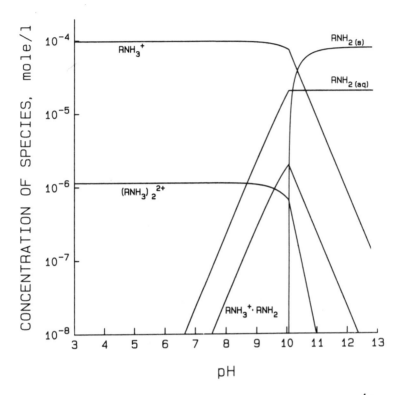

FIG. 10   Logarithmic concentration diagram for $1 \times 10^{-4}$ M total dodecylamine.

Iwasaki et al. (48) found complete flotation to take place between pH 4 and 12.  These results clearly show the close relation between the PZC of the mineral (pH 2 for quartz) and its flotation behavior with a cationic collector.

Oxide minerals such as hematite, goethite, magnetite, and corundum also float well with cationic collectors above their points of zero charge.  The results of quartz flotation with amines (and alkyl sulfonates) have been reported by Iwasaki et al. (49) and of corundum with amines (and sulfonates) by Modi and Fuerstenau (50).  The flotation response of these two minerals with dodecylamine salts and sodium dodecyl sulfonates is reproduced in Fig. 11.  Figure 11 also presents the $\zeta$ potential of quartz and corundum, clearly showing at which

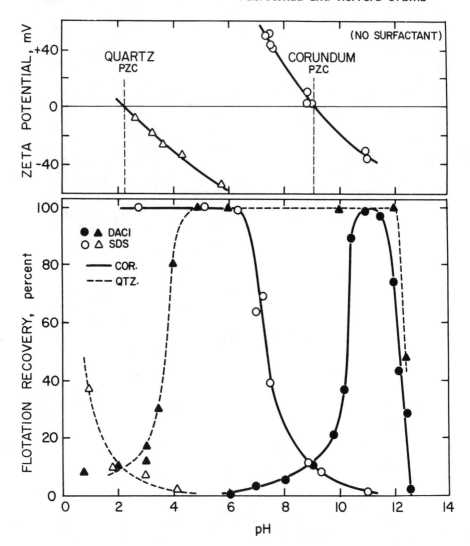

FIG. 11    Correlation of the ζ potential as a function of pH
for quartz and corundum with flotation recovery with
$4 \times 10^{-5}$ M sodium dodecyl sulfonate and dodecylammonium
chloride as collector.

pH the minerals reverse their surface charge. As expected, the flotation of quartz with the amine is optimum above its PZC, while maximum corundum flotation occurs between pH 10 and 12, since its PZC corresponds to pH 9. For both minerals, flotation ceases at pH 12.5 because the collector hydrolyzes at this high pH and the concentration of aminium ions in solution is consequently very small. Figure 11 clearly illustrates the strong dependence of quartz and corundum flotation with cationic collectors on the PZC of the mineral. It also indicates that the ionized form of the surfactant is the species acting as the collector and that in the pH range 4–9 quartz can be separated from corundum. As will be discussed in the next section, the converse is true for flotation with the anionic collector. Figure 11 clearly shows how the PZC of minerals controls flotation with physically adsorbing cationic and anionic surfactants. Figure 12 summarizes the cationic flotation response of three different oxides, namely, corundum, ilmenite, and quartz, relative to their PZC values. The flotation behavior of these oxides with dodecylamine as the collector seems to be controlled mainly by electrostatic interactions with the mineral surface.

Cases (51) investigated the flotation behavior of a series of silicate minerals with dodecylamine as the collector and found physical forces to dominate the modes of adsorption of the collector. Spodumene flotation with dodecylammonium chloride also indicates a flotation behavior typical of a system involving electrostatic attraction between a negatively charged silicate and a cationic collector (30). The flotation of aluminosilicate minerals with cationics has been conducted by Choi and Oh (52). In the case of kyanite ($pH_{PZC} = 6.9$), they found that an appreciable amount of the collector adsorbs below the PZC. This observation points out the dissolution problem common to complex silicate minerals. If this type of silicate is immersed in aqueous solutions, the metal cations constituting its crystal lattice will dissolve preferentially, thus reducing the metal/Si ratio and lowering the apparent PZC. The cationic flotation of oxides and silicates has been reviewed by Smith And Akhtar (53).

The results of several investigations on the cationic flotation of such sparingly soluble minerals as calcite (54,55) and apatite (54) have been reported. The flotation behavior of these minerals with dodecylamine as the collector confirms that the predominant mode of adsorption of this cationic on both calcite and apatite is due to Coulombic and van der Waals forces.

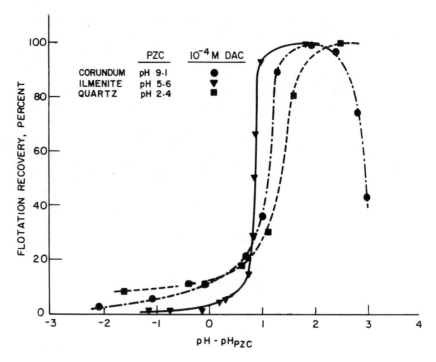

FIG. 12    Flotation response of a number of oxide minerals as
a function of the pH relative to the pH of their PZC in the
presence of $10^{-4}$ M dodecylammonium acetate.

Hemimicelle formation and the coadsorption of the neutral amine
molecule with the charged species also seem to aid the flotation
process in these systems.

     Such soluble salts as sylvite (KCl) also respond to flotation
with amines.  It floats with 12- and 14-carbon amines only after
precipitation of amine chloride (56).  When octylamine is used
as the collector, however, sylvite floats before amine chloride
precipitation, but relatively high concentrations of the collector
are required for complete flotation.  Figure 13 clearly indicates
the conditions for effecting the separation of sylvite from
halite (NaCl) by flotation using octylamine as the collector.
Flotation systems involving soluble salts exhibit ionic strengths
that are on the order of 5 M.  This high ionic strength has a
big impact on both the $\zeta$ potential and the solubility of the

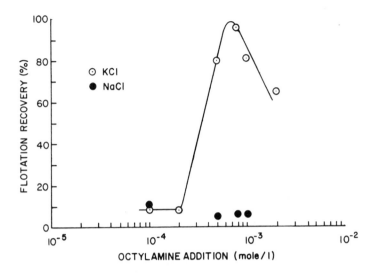

FIG. 13   Flotation recovery of KCl and NaCl as a function of octylamine addition.

collector.  Because of the collapse of the electrical double layer under these conditions, the $\zeta$ potential is zero and the thickness of the double layer is equivalent to an ionic size. Fuerstenau and Fuerstenau (57) suggested that the mechanism of adsorption of amines on KCl involves an ion exchange process. Roman et al. (58) later postulated that amine cations may possibly adsorb in vacant $K^+$ sites and neutral molecules in vacant $Cl^-$ sites.

2.   Mineral—Anionic Collector Systems

Anionic surfactants that adsorb on mineral surfaces through electrostatic attraction and that are commonly used as collectors in froth flotation belong to four groups of chemical reagents: alkyl carboxylates, alkyl sulfates, alkyl sulfonates, and alkyl phosphates.  With the exception of alkyl sulfonates, these anionic compounds hydrolyze in acidic solutions and they can consequently function as collectors only above their $pK_a$ values. Similar to the case for amines, the pH controls the solution chemistry of alkyl carhoxylates, phosphates, and, to a lesser extent, sulfates, and it will thus determine the conditions of the

solution where the reagent can act as an anionic collector.
The $pK_a$ values of long-chain carboxylates are in the pH range
4–10 (59), while those of alkyl sulfates occur at more acidic
pH values since they are completely dissociated above pH 3
(60). Because alkyl sulfonates are not hydrolyzed even by
strong acids, they are completely ionized in aqueous solutions
and the pH does not affect their ability to function as anionic
collectors.

Dodecyl sulfate anions have been shown to adsorb physically
onto corundum (50). In this system, flotation occurs only at
pH values at which the surface of the mineral is positively
charged, namely, at pH 4 and 6, as demonstrated by the
results presented in Fig. 14. Similarly, both dodecyl sulfate
and sulfonate anions are the active collecting species in the
dodecyl sulfate- and dodecyl sulfonate-goethite systems (61),
and the dodecyl sulfate-hematite system (49). Iron oxide flotation in these systems occurs only at pH values below the PZC
of the mineral indicating an adsorption mechanism involving
mainly physical forces. The flotation results presented in
Fig. 15 also indicate that physical forces are mainly involved
in the adsorption of dodecyl sulfonate on manganese dioxide
because flotation occurs only below pH 6, which corresponds
to the PZC of this mineral (62). Under these conditions, the
mineral surface is positively charged and hence adsorption of
the anionic sulfonate is by electrostatic attraction. Relatively
short-chain carboxylates adsorb physically by electrostatic
interaction with the mineral surface of oxides. Flotation of
chromite in the presence of $10^{-4}$ M sodium laurate, for example,
is possible only below its PZC (63). Flotation of silicates
with anionic collectors that adsorb physically on their surfaces
is also determined by the PZC of the mineral. The results of
flotation experiments with silicates and dodecyl sulfate as the
collector confirm that the flotation response of these minerals
is controlled by their PZC (51,52,64).

## B.  Flotation by Chemical Adsorption of the Collector

Collectors that adsorb on the surface of minerals through
chemical forces provide the means for effecting flotation
selectivity. Chemical interaction between the polar headgroup
of the collector and the anions or cations constituting the surface
sites involves charge transfer that leads to the formation of a
covalent bond. Therefore, when ionic collectors chemisorb,
they can adsorb even on highly charged mineral surfaces that

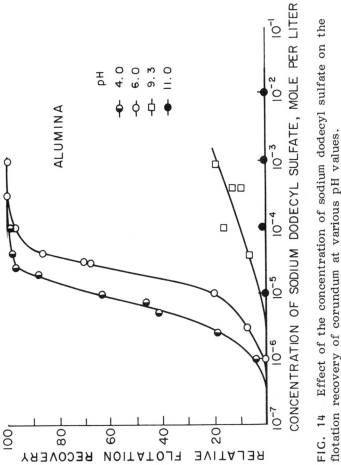

FIG. 14   Effect of the concentration of sodium dodecyl sulfate on the flotation recovery of corundum at various pH values.

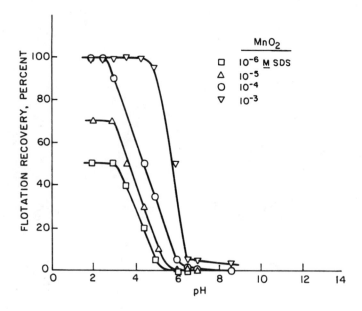

FIG. 15    Flotation response of manganese dioxide as a function of pH with different concentrations of sodium dodecyl sulfonate.

carry a charge similar to that of the ion.  When chemically adsorbing surfactants are used as collectors, the flotation behavior of the mineral is not related to its PZC.  This was clearly established by Iwasaki et al. (65), who investigated the flotation response of hematite as a function of pH using three unsaturated fatty acids that are known to chemically react with iron.  Although the PZC of this hematite sample occurs at pH 6.7, its flotation with these anionic collectors is possible up to about pH 9 with linolenic, pH 10 with linoleic, and pH 11 with oleic acid.  In these systems, chemisorption of the collector must be occurring; otherwise adsorption could not take place under conditions where the surface potential is highly negative.  As already shown by the adsorption isotherms presented in Fig. 6 for the oleate-hematite system, oleate anions adsorb on hematite even when its surface is negatively charged.  Under these conditions the adsorption process involves a chemical mechanism that results in the formation of ferric oleate.

The process of chemical interaction between mineral and collectors becomes more complicated if the ions that constitute the crystal lattice have a tendency to dissolve and form collector complexes in solution. Complex formation may lead to bulk precipitation of metal salts and anion-collector compounds. Surfactants that function as flotation collectors through the formation of chemical species with the ions that constitute the mineral lattice are mainly fatty acids, alkyl sulfonates, alkyl sulfates, and chelating agents. Amines have been reported to chemically interact with apatite by reacting with phosphate ions (66).

Fatty acids have been found to function as flotation collectors for oxide, silicate, and sparingly soluble minerals by adsorbing chemically at the mineral-aqueous solution interface. The flotation chemistry of such sparingly soluble minerals as apatite and dolomite with oleate as the collector has been recently investigated by Deason (67). The effective flotation separation of the phosphate-bearing mineral apatite from carbonates is vital to beneficiate dolomitic phosphate ores. This separation, however, is made difficult by the similar nature of their surfaces and the presence of substantial quantities of dissolved species. Comparison of oleate flotation of dolomite and apatite following individual and mutual equilibration is shown in Fig. 16. These results clearly indicate that in both cases the flotation response of these minerals is not affected by mutual equilibration. In the case of dolomite, even though equilibration with apatite below pH 9 results in phase transformation at the surface, its flotation properties do not change. Clearly, separation of dolomite from apatite by oleate flotation can be achieved in acidic slurries. In the alkaline pH region, oleate adsorbs chemically on both apatite and dolomite.

Those reagents that are known to form chelates with metal ions in the mineral lattice have been shown to be the most promising flotation collectors. Alkyl hydroxamates, oximes, oxines, alkyl aryl phosphonic acids, alkane dicarboxylic acids, and alkyl arsonic acids are among the chelating agents that have been shown to be effective collectors for the flotation of minerals. Both zinc and lead oxide minerals, for example, respond to flotation when 8-hydroxyquinoline is used as the collector (68). In these systems the oxine strongly reacts with the cations at the mineral surface and forms an insoluble, highly stable metal chelate. The chemistry and utilization of chelating agents in mineral flotation has been recently reviewed by Somasundaran (69). Hydrophilic organic reagents that are known to chelate metal ions have found application as flotation depressants. The selective flotation separation of cassiterite

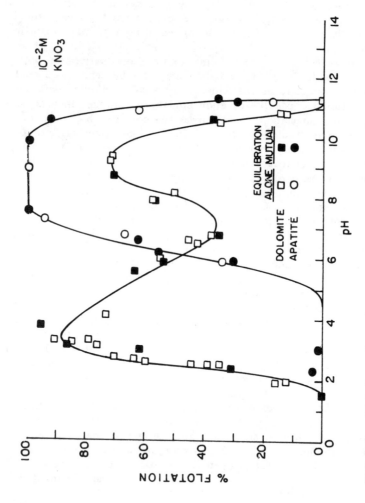

FIG. 16   Flotation of apatite and dolomite as a function of pH with sodium oleate following individual and mutual equilibration.

($SnO_2$) from tin ores using alkane dicarboxylic acid as collector
is accomplished only when aminonaphthol sulfonic acids are
used to depress topaz, an aluminosilicate mineral that also
responds to flotation with dicarboxylic acids (70). The formation on topaz of an aluminoaminonaphthol sulfonate chelate
through a N→Al-O bond may be responsible for the preferential
adsorption of the depressant over the collector species.

The potassium salt of octyl hydroxamic acid, a weak electrolyte whose $pK_a$ value corresponds to pH 9.7 (71), has been
widely investigated as a potential collector in oxide, silicate,
and sparingly soluble mineral flotation. Flotation of manganese
dioxide with this chelating agent as the collector has clearly
established the chemical nature of the adsorption mechanism
(62). Hydroxamate flotation of manganese dioxide takes place
at pH values where the surface of the mineral is highly negatively charged and where manganese-hydroxamate complexes are
stable. The PZC of this manganese dioxide, $\gamma$-$MnO_2$, occurs
at pH 5.6, while the pH of maximum flotability is about pH 9.
This maximum in mineral flotation with hydroxamate has also
been reported for hematite and rhodonite (8) and chrysocolla
(72). The formation of iron, copper, and manganese hydroxamate chelates on the surface of minerals containing these metals
has been confirmed by infrared spectroscopy.

In a recent study in an attempt to improve the separation
of bastnaesite [($La,Ce)CO_3F$] from barite and calcite, it was
found that conditions can be achieved in which bastnaesite can
be floated quite effectively with hydroxamate as the collector
(73). This probably results because hydroxamate chelates
more strongly with rare-earth metal ions than with $Ba^{2+}$ and
$Ca^{2+}$. The hydroxamate flotation response of these three
minerals at different pH values is given in Fig. 17. From these
results it is clearly seen that in neutral or mild alkaline solutions only bastnaesite responds to flotation with octyl hydroxamate, while both calcite and barite flotation recovery is essentially zero. Fuerstenau and Pradip (73) summarized the published results on the hydroxamate adsorption and flotation
response of various minerals, and these are given in Table 4.
These data show that, with the exception of chrysocolla,
pyrochlore, and quartz-microcline, the minerals investigated
(with PZC values ranging from pH 2 to 10) float best at
around pH 9 when hydroxamate is used as the collector.
Table 4 also gives the pH of maximum hydroxamate adsorption
on some minerals. Since this pH is about 9 in all cases, these
results indicate that the extent of collector adsorption at the

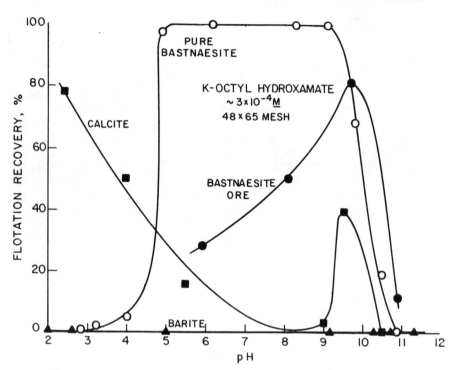

FIG. 17    Flotation response of calcite, barite, and bastnaesite as a function of pH with $3 \times 10^{-4}$ M octyl hydroxamate as the collector.

solid-liquid interface might be used to predict the hydroxamate flotation behavior of minerals.

## C.  Flotation Involving Activation Phenomena

Activation phenomena refer to the modification of the electrical nature of mineral-aqueous solution interfaces brought upon after the adsorption in the Stern layer of ions oppositely charged to the surface.  By means of this adsorption process, minerals that naturally exhibit a positive or negative charge can be made to respond to flotation with cationic or anionic collectors, respectively.  In flotation terminology, those inorganic ions that exhibit surface activity and are able to modify the electrical nature of the interface are called activators.  Mineral activation is

TABLE 4   Summary of Studies on the Flotation of Various Minerals with Hydroxamate Collectors

| Mineral | Chemical formula | pzc (pH) | Reagent | Method of investigation | pH of optimum flotation or maximum adsorption |
|---|---|---|---|---|---|
| Hematite | $Fe_2O_3$ | 8.2 | K-octyl hydroxamate | Flotation | 9.0 |
| Hematite | $Fe_2O_3$ | 8.2 | K-octyl hydroxamate | Oil extraction microflotation, contact angles with air/oil | 8.0–8.5 |
| Hematite | $Fe_2O_3$ | 8.2 | K-octyl hydroxamate | Adsorption | 8.5 |
|  | $\gamma-MnO_2$ | 5.6 | K-octyl hydroxamate | Flotation | 9.0 |
|  | $\gamma-MnO_2$ | 5.6 | K-octyl hydroxamate | Adsorption | 9.0 |
| Rhodonite | $(mN,Fe,Cu)SiO_3$ | 2.8 | K-octyl hydroxamate | Flotation | 9.0 |
| Chrysocolla | $CuSiO_3 \cdot 2H_2O$ | 2.0 | K-octyl hydroxamate | Flotation | 6.0 |
| Chrysocolla | $CuSiO_3 \cdot 2H_2O$ | 2.0 | K-octyl hydroxamate | Flotation | 6.0 |

**TABLE 4** (Continued)

| Mineral | Chemical formula | pzc (pH) | Reagent | Method of investigation | pH of optimum flotation or maximum adsorption |
|---|---|---|---|---|---|
| Malachite | $Cu_2CO_3(OH)_2$ | 7.9 | K-octyl hydroxamate | Flotation | 6–10; 9.5 at $10^{-4}$ M |
| Malachite | $Cu_2CO_3(OH)_2$ | 7.9 | K-octyl hydroxamate | Adsorption | 9, adsorption below 5 |
| Chrysocolla/ malachite/ azurite | | — | C6–C9 hydroxamate | Flotation | 6.5–9.5 for good flotation; best in plant at 7.5–8.0 |
| Pyrochlore | $NaCaNb_2F(CO_3)_6$ | — | IM-50 (C7–C9) | Flotation | 6.0 |
| Fluorite | $CaF_2$ | — | IM-50 | Flotation | 8.5 |
| | | | | Adsorption | 8.5 |

| Mineral | Formula | | Reagent | Process | pH |
|---|---|---|---|---|---|
| Huebnerite | $MnWO_4$ | — | IM-50 | Flotation | 9.0 |
| | | | | Adsorption | 9.0 |
| Barite | $BaSO_4$ | 10 | K-octyl hydroxamate | Flotation | 9.5 |
| | | | | Adsorption | 9.0 |
| Calcite | $CaCO_3$ | 10 | K-octyl hydroxamate | Flotation | 9.5, also recovery below 8 |
| | | | | Adsorption | 9.5, also adsorption below 8 |
| Bastnaesite | $(Ce,La)FCO_3$ | 9.5 | K-octyl hydroxamate | Flotation | 5–9 |
| | | | | Adsorption | 7–8.5 |
| Quartz/microline | $SiO_2$ | 2.0 | K-octyl hydroxamate | Flotation | 1.5 |
| Oxidized Zn-Pb ores | — | — | C6,8 hydroxamate | Flotation | — |

usually achieved upon the addition of certain chemical reagents
into mineral suspensions. In some cases, however, the release
into solution of metal cations that constitute the crystal lattice
of minerals may lead to surface activation. Under certain pH
conditions, these metal ions hydrolyze and readsorb at the
mineral-water interface thus reversing the sign of the $\zeta$ potential. This phenomenon is known as autoactivation.

Metal hydroxo complexes are among the most common cations
that function as activators in mineral flotation. Figure 18
shows the sulfonate flotation response of quartz activated with
a variety of metal cations (7). Only the initial flotation edge
is shown for clarity and to illustrate the close correlation
between flotation and metal ion hydrolysis. The pH range of
maximum flotation in these systems is the same pH region in
which the metal hydroxo complexes predominate in solution.
Since the PZC of quartz occurs at about pH 2, it responds to
flotation with a sulfonate or a fatty acid only in the presence
of an activating metal ion. Various complex oxide and silicate
minerals have been found to respond to flotation at several different pH values. Because of the slight solubility of these
minerals, their flotation behavior is probably associated with
autoactivation by the hydrolyzed species of the cations constituting the mineral lattice. The oleate flotation of chromite, an
oxide mineral containing both ferrous iron and magnesium,
exhibits pronounced peaks at the pH of maximum ferrous and
magnesium hydroxo complexes concentration (63). A similar
flotation behavior has been reported for spodumene when oleate
flotation response of this complex silicate mineral (whose PZC
occurs at pH 2.5) as a function of pH. Flotation peaks at
two pH values. The major peak at about pH 8 appears to be
chemisorption of oleate on aluminum sites, i.e., on the sites of
the cations that are holding the silicate chains together. Maximum flotation at pH 4, on the other hand, appears to result
from autoactivation of the mineral upon dissolution of lattice
ferric ions and readsorption of ferric hydroxo complexes. The
role of mineral autoactivation in flotation is clearly seen by
comparing the flotation response of a low- and a high-iron
spodumene shown in Fig. 19. The increase in flotation of the
high-iron spodumene sample indicates that hydrolysis of ferric
ions might aid the complex spodumene collection process.

Activation of mineral surfaces can also be achieved upon
the adsorption of certain anions at the mineral-aqueous solution
interface. For example, fluoride has been widely used as an
activator in the cationic flotation separation of feldspar from

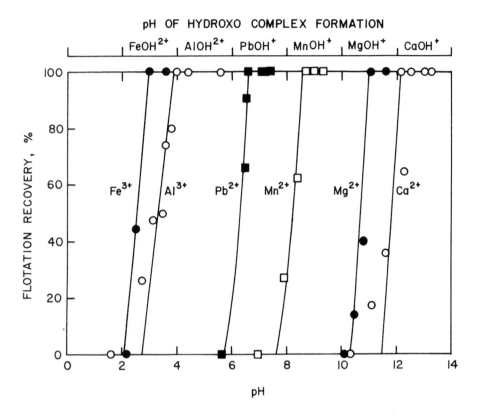

FIG. 18   Minimum flotation edges of quartz as a function of
pH with $1 \times 10^{-4}$ M metal ions as activators and $1 \times 10^{-4}$ M
dodecyl sulfonate as collector.


quartz, and as a depressant in the anionic flotation of spodumene and beryl (74).   Smith (75) assessed the effect of fluoride
addition on the wettability of microcline feldspar and quartz in
the presence of aqueous dodecylamine solutions by measuring
the contact angle at different pH values in the absence
and presence of sodium fluoride.   In the absence of fluoride,
quartz and microcline exhibit an identical wetting behavior in
the presence of the amine collector.   While contact angles on
quartz are not affected by the addition of fluoride, those on
microcline change significantly, especially at low pH values,
upon fluoride activation.   To improve the processing of ilmenite

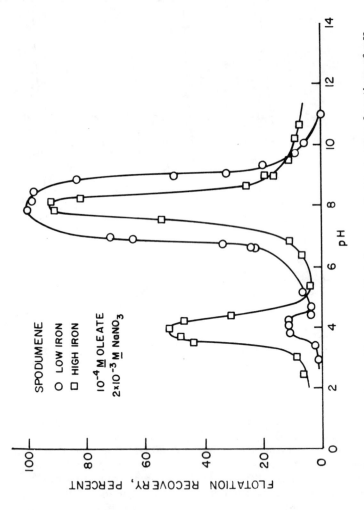

FIG. 19  Flotation of low- and high-iron spodumene as a function of pH in the presence of $1 \times 10^{-4}$ M oleate. (From Ref. 30.)

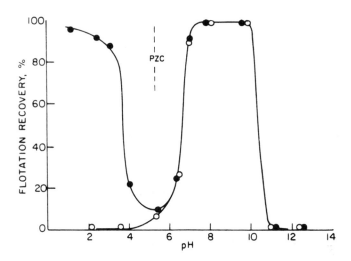

FIG. 20    The effect of hydrochloric acid (○) and sulfuric acid
(●) on the flotation of ilmenite with dodecylammonium acetate
as the collector.

sands, Nakatsuka et al. (76) investigated the flotation of hematite, ilmenite, magnetite, quartz, and feldspar using HCl and
$H_2SO_4$ as the pH regulator.    Their flotation results for ilmenite
($FeTiO_3$) are presented in Fig. 20, which shows that ilmenite
flotation with dodecylammonium acetate as the collector at pH
values below its PZC is achieved only when sulfate ions are
used as activators.    In the presence of either HCl or $H_2SO_4$, as
the pH is lowered toward the PZC, namely, pH 5.2, the flotation
of ilmenite decreases in the expected manner.    In the presence
of $H_2SO_4$, however, flotation rises sharply again at pH values
below the PZC because sulfate anions activate ilmenite for
cationic flotation.

## VII.    FLOTATION OF FINE-SIZED MINERAL PARTICLES

An increasingly common problem encountered in mineral flotation
is the recovery of fine-sized particles (less than 10 μm) containing valuable mineral resources.    These particles are generated
in the mineral industry during the comminution stage because
of its low efficiency and, if necessitated, by liberation

requirements as in the processing of low-grade ore bodies that
contain finely disseminated minerals. Several techniques have
been proposed to increase mineral recovery by flotation in
plants treating low-grade, fine-grained ores (77). One method
exhibiting significant potential for recovering valuable minerals
from fine-particle suspensions involves selective flocculation
before froth flotation is used. Polymers of long-chain organic
molecules are used to selectively aggregate certain mineral con-
stituents into "flocs." Depending on the process and mineral
composition, either the valuable or the gangue mineral may be
flocculated. To obtain a suitable concentrate, the flocculated
particles may be separated from the suspension by froth flotation. Such a process is termed floc flotation.

## A.  Floc Flotation

Fuerstenau and Gebhardt (78) investigated the floc flotation of
hematite fines using polyacrylic acid (PAA) as the flocculant
and sodium dodecyl sulfonate (SDS) as the collector. Their
results are presented in Fig. 21, where the amount of hematite
floated at pH 4.1 is plotted as a function of SDS adsorption
density in the absence and presence of various preadsorbed
amounts of polyacrylic acid. Optimum flotation of the primary
hematite particles without polymer requires relatively high SDS
adsorption densities. Low polymer dosages before collector
addition aid flotation by forming flocs of primary particles thus
reducing the total surface area available for collector adsorption. Excess polymer, however, results in low flotation yields
due to surface saturation by the hydrophilic polymer.

## B.  Carrier Flotation

Separation of fine-sized mineral particles by froth flotation is
also achieved by finding conditions under which these fine
particles will attach to the surface of large size particles that
act as the carrier mineral, finding a suitable collector and
frother, and floating the agglomerates. This process is known
as carrier flotation or ultraflotation, and its first commercial
application was in the purification of kaolin at Minerals and
Chemical Phillipp's plant at McIntyre, Georgia (77). Carrier
flotation of hematite fines using prereagentized coarse hematite
particles as the carrier was recently investigated by Li (79),

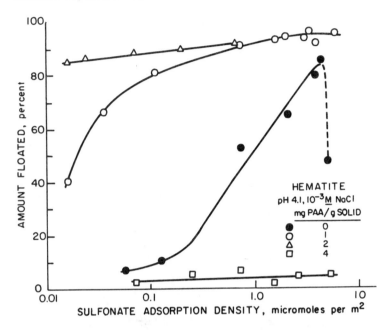

FIG. 21    Amount of hematite floated at pH 4.1 as a function of sodium dodecyl sulfonate adsorption density in the absence and presence of various preadsorbed amounts of polyacrylic acid.

and his results are reproduced in Fig. 22.    Separation of hematite aggregates from unflocculated quartz fine particles was achieved by flotation with sodium dodecyl sulfate and improved by adding prereagentized coarse hematite particles during the shear flocculation treatment.    The use of a certain amount of prereagentized coarse hematite in the flocculation treatment improved the separation performance, probably due to carrier effects as related not only to the particles but also to the reagent.    The carrier flotation results given in Fig. 22 show that increasing the amount of prereagentized coarse particles from 0 to 40% by weight relative to the hematite fines significantly improves the recovery as well as increases the selectivity.

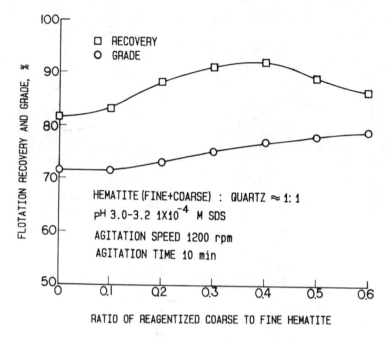

FIG. 22    Effect of the amount of reagentized coarse hematite on the flotation recovery and grade of fine hematite.

## VIII.  SUMMARY

Surface chemistry is the very basis of froth flotation.  Judicious application of surface chemistry (physical adsorption and chemical adsorption) plus crystal chemistry can be used to effect mineral separation by flotation.  An interesting way to summarize this is to illustrate the flotation separation of a series of silicate minerals, namely, talc, spodumene, muscovite (mica), feldspar, and quartz from pegmatitic ores.  Chemically similar but structurally different, these silicates can be separated from each other by flotation.  Such flotation separations are possible because silicate minerals exhibit widely varying crystallographic structures, depending on the manner in which silicon-oxygen tetrahedra polymerize and any isomorphous substitution in the lattice.  Because of these differences in crystal structure, silicates range from having a hydrophobic surface, a fixed surface charge, extensive metal surface sites, to having

TABLE 5    Selective Flotation Separation of Silicate Minerals in
Pegmatitic Ores

| Silicate mineral | Chemical formula | pzc (pH) | Flotation pH | Flotation reagents added |
|---|---|---|---|---|
| Talc | $Mg_3(Si_4O_{10})(OH)_2$ | 3.6 | 6 | MIBC frother |
| Spodumene | $LiAl(Si_2O_6)$ | 2.5 | 8 | Oleate as collector |
| Muscovite | $KAl_2(AlSi_3O_{10})(OH)_2$ | 3 | 3 | Amine as collector |
| Feldspar | $K(AlSi_3O_8)$ | 2-3 | 2.5 | HF as activator, amine as collector |
| Quartz[a] | $SiO_2$ | | | |

[a]In such an ore, the final product is essentially pure quartz since all
of the other minerals would have been recovered successively by
selective flotation as outlined.

broken Si-O bonds that control surface behavior.  Since the
surfaces of silicates are generally charged because of broken
Si-O bonds at the surface, electrostatic adsorption of collector
ions is important in the flotation of such minerals.  In the case
of chemisorbing collectors, chemical interaction of the collector
with metal sites on certain silicates provides the means for effecting flotation selectivity.

Table 5 summarizes the conditions for the selective separation of a series of silicate minerals in a pegmatitic ore by froth
flotation.  If talc were present, use can be made of its natural
hydrophobicity to separate it from the other siliceous minerals.
To do this, only a frother need be added to the system.  Under
these conditions talc can be floated while the other silicates
remain in the flotation cell.  After talc removal, the second step
involves spodumene flotation at pH 8 with oleate.  Only chemisorption of oleate permits simple flotation separation of spodumene from other silicate minerals.  After spodumene has been
removed, lowering the pH to about 3 yields conditions for
muscovite flotation with a cationic collector.  Under these conditions quartz and feldspar do not respond to flotation with

amines. Since muscovite is a layer silicate with a lattice
charge, it retains a negative surface charge on the faces of
the crystal sheets, where the cationic amine adsorbs. Subsequently, feldspar can be separated from quartz by flotation
upon feldspar activation with fluoride and flotation at pH 2–3
with a cationic collector. Since the surfactants used as collectors also act as foaming agents, generally little or no additional
frother need be added.

## REFERENCES

1.  D. L. Edelstein and G. H. Hyde, in *Bureau of Mines
    Minerals Yearbook*, U.S. Department of the Interior,
    Washington, D.C., 1985.
2.  *Mineral Industry Surveys, Froth Flotation in the United
    States*, Bureau of Mines, Washington, D.C., 1987, p. 33.
3.  D. W. Fuerstenau and T. W. Healy, in *Adsorptive Bubble
    Separation Techniques*, Academic Press, New York, 1972,
    Chap. 6.
4.  J. Leja and B. Q. He, in *Principles of Mineral Flotation*
    (M. H. Jones and J. T. Woodcock, eds.), Australasian
    Institute of Mining and Metallurgy, Victoria, Australia,
    1984, p. 73.
5.  G. A. Parks, *Adv. Chem. Ser.*, 67: 121 (1967).
6.  M. C. Fuerstenau and B. R. Palmer, in *Flotation, A. M.
    Gaudin Memorial Volume* (M. C. Fuerstenau, ed.), Vol. 1,
    AIME, New York, 1976, p. 148.
7.  H. S. Hanna and P. Somasundaran, in *Flotation, A. M.
    Gaudin Memorial Volume* (M. C. Fuerstenau, ed.), Vol. 1,
    AIME, New York, 1976, p. 197.
8.  P. Somasundaran and G. Agar, *J. Colloid Interf. Sci.*,
    24: 433 (1967).
9.  D. W. Fuerstenau and S. Raghavan, in *Proceedings XII
    Int. Minerals Processing Congr.*, Vol. 2, Sao Paulo,
    Brazil, 1980, p. 480.
10. J. T. Davies and E. K. Rideal, *Interfacial Phenomena*,
    2nd ed., Academic Press, New York, 1963, p. 480.
11. R. J. Hunter, *Zeta Potential in Colloid Chemistry*,
    Academic Press, New York, 1981, p. 386.
12. P. H. Wiersema, A. L. Loeb, and J. T. G. Overbeek,
    *J. Colloid Interf. Sci.*, 22: 78 (1966).

13. D. W. Fuerstenau and S. Chander, in *Industrial Applications of Surface Analysis* (L. A. Casper and C. J. Powell, eds.), ACS Symp. Ser., Vol. 199, 1982, p. 199.

14. D. W. Fuerstenau, in *Physics and Chemistry of Porous Media* (D. L. Johnson and P. N. Sen, eds.), American Institute of Physics, New York, 1984, p. 209.

15. A. M. Schwartz, in *Wetting, Spreading, and Adhesion* (J. F. Padday, ed.), Academic Press, New York, 1978, p. 93.

16. F. M. Fowkes, in *Wetting S.C.I. Monograph No. 25*, Society of Chemistry and Industry, London, 1967.

17. F. M. Fowkes, in *Contact Angle, Wettability and Adhesion* (R. F. Gould, ed.), Adv. Chem. Ser., Vol. 43, 1964, p. 99.

18. J. Laskwoski and J. A. Kitchener, *J. Colloid Interf. Sci.*, 29: 670 (1969).

19. K. L. Sutherland and I. W. Wark, *Principles of Flotation*, Australasian Institute of Mining and Metallurgy, Melbourne, Australia, 1955.

20. C. A. Smolders, *Rec. Trav. Chem.*, 80: 651 (1961).

21. P. L. deBruyn, J. T. Overbeek, and R. Schuhmann, Jr., *Trans. AIME, 199*: 519 (1954).

22. F. F. Aplan and P. L. deBruyn, *Trans. AIME, 226*: 235 (1963).

23. P. Somasundaran, *Trans. AIME, 241*: 105 (1968).

24. D. W. Fuerstenau and S. Raghavan, in *Flotation, A. M. Gaudin Memorial Volume* (M. C. Fuerstenau, ed.), Vol. 1, AIME, New York, 1976, p. 21.

25. D. W. Fuerstenau, in *Principles of Mineral Flotation* (M. H. Jones and J. T. Woodcock, eds.), Australasian Institute of Mining and Metallurgy, Victoria, Australia, 1984, p. 7.

26. D. W. Fuerstenau, Sc.D. thesis, Massachusetts Institute of Technology, 1953.

27. A. M. Gaudin and D. W. Fuerstenau, *Min. Eng.*, 7: 958 (1955).

28. D. W. Fuerstenau, *J. Phys. Chem.*, 60: 981 (1956).

29. D. W. Fuerstenau, T. W. Healy, and P. Somasundaran, *Trans. AIME, 229*: 321 (1964).

30. K. S. Moon and D. W. Fuerstenau, AIME Annual Meeting, Chicago, 1981.

31. K. P. Ananthapadmanabhan, P. Somasundaran, and T. W. Healy, *Trans. AIME, 266*: 2003 (1979).

32.  D. W. Fuerstenau, *Min. Eng.*, *9*: 1365 (1957).
33.  R. Herrera-Urbina, Ph.D. thesis, University of California, Berkeley, 1985.
34.  J. Laskowski and J. Iskra, *Trans. IMM*, Sec. C, *79*: 6 (1970).
35.  B. V. Derjaguin and S. S. Dukhin, *Trans. IMM*, *70*: 221 (1961).
36.  K. L. Sutherland, *J. Phys. Chem.*, *52*: 394 (1984).
37.  R. L. Flint and W. I. Howarth, *Chem. Eng. Sci.*, *26*: 1155 (1971).
38.  V. I, Klassen and V. A. Mokrousov, *An Introduction to the Theory of Flotation*, Buttersworths, London, 1963, p. 493.
39.  B. V. Derjaguin and M. M. Kusakov, *Acta Physicochemica USSR*, *10*: 153 (1939).
40.  I. C. Callaghan and K. W. Baldry, in *Wetting, Spreading, and Adhesion* (J. F. Padday, ed.), Academic Press, New York, 1978, p. 161.
41.  B. V. Derjaguin, Z. M. Zorin, N. V. Churaev, and V. A. Shishin, in *Wetting, Spreading, and Adhesion* (J. F. Padday, ed.), Academic Press, New York, 1978, p. 201.
42.  B. V. Derjaguin and S. S. Dukhin, in *Mineral Processing: Developments in Mineral Processing*, Vol. 2 (J. Laskowski, ed.), Elsevier, Amsterdam, 1981, p. 21.
43.  B. V. Derjaguin and N. S. Shukakidse, *Trans. IMM*, *70*: 569 (1961).
44.  A. Scheludko, *Kolloid Z. Z. Polym.*, *191*: 52 (1963).
45.  J. Lekki and J. Laskowski, *Trans. IMM*, Sec. C, *80*: 174 (1971).
46.  J. L. Yordan and R. H. Yoon, AIME Annual Meeting, New Orleans, 1985.
47.  S. H. Castro, R. M. Vurdela, and J. S. Laskowski, *Colloid Surf.*, *21*: 87 (1986).
48.  I. Iwasaki, S. R. B. Cooke, D. H. Harraway, and H. S. Choi, *Trans. AIME*, *223*: 97 (1962).
49.  I. Iwasaki, S. R. B. Cooke, and H. S. Choi, *Trans. AIME*, *223*: 97 (1962).
50.  H. J. Modi and D. W. Fuerstenau, *Trans. AIME*, *217*: 381 (1960).
51.  J. M. Cases, *Trans. AIME*, *247*: 123 (1970).
52.  H. S. Choi and J. Oh, *J. Inst. Min. Metall. Japan*, *81*: 614 (1979).

53. R. W. Smith and S. Akhtar, in *Flotation, A. M. Gaudin Memorial Volume*, Vol. 1 (M. C. Fuerstenau, ed.), AIME, New York, 1976, p. 87.

54. S. K. Mishra, *Int. J. Min. Process.*, *6*: 119 (1979).

55. R. E. Arnold, E. E. Brownbill, and S. W. Ihle, *Int. J. Min. Process*, *5*: 143 (1978).

56. M. C. Fuerstenau, J. D. Miller, and M. C. Kuhn, *Chemistry of Flotation*, SME-AIME, New York, 1985, p. 177.

57. D. W. Fuerstenau and M. C. Fuerstenau, *Min. Eng.*, *8*: 302 (1956).

58. R. J. Roman, D. C. Seidel, and M. C. Fuerstenau, *Trans. AIME*, *241*: 56 (1968).

59. C. du Rietz, *Proceedings 2nd Scandinavian Symposium on Surface Activity*, Stockholm, Academic Press, New York, 1964, p. 21.

60. J. Leja, *Surface Chemistry of Froth Flotation*, Plenum Press, New York, 1982, p. 758.

61. I. Iwasaki, S. R. B. Cooke, and A. F. Colombo, RI No. 5593, U.S. Bureau of Mines, Washington, D.C., 1960.

62. R. Natarjan and D. W. Fuerstenau, *Int. J. Min. Process.*, *11*: 139 (1983).

63. B. R. Palmer, M. C. Fuerstenau, and F. F. Aplan, *Trans. AIME*, *258*: 261 (1975).

64. I. Iwasaki, S. R. Cooke, and H. S. Choi, *Trans. AIME*, *220*: 394 (1961).

65. I. Iwasaki, S. R. Cooke, and H. S. Choi, *Trans. SME*, *217*: 237 (1960).

66. H. Soto and I. Iwasaki, *Min. Met. Process.*, *2*: 160 (1985).

67. D. M. Deason, Ph.D. thesis, University of California, Berkeley, 1987.

68. G. Rinelli and A. M. Marabini, *Proc. Xth International Minerals Processing Congress*, Cagliary, Italy, 1973, p. 493.

69. P. Somasundaran, in *Reagents in the Minerals Industry* (M. J. Jones and R. Oblatt, eds.), IMM, London, 1984, p. 209.

70. H. Baldauf, J. Schoenherr, and H. Schubert, *Int. J. Min. Process.*, *15*: 117 (1985).

71. V. I. Ryaboi, V. A. Shenderovich, and E. F. Strizhev, *Russian J. Phys. Chem.*, *54*: 730 (1980).

72. D. W. Fuerstenau, R. Herrera-Urbina, and J. Laskowski, in *Proc. II Latin American Congress of Froth Flotation*, Vol. II, Chile, 1985, p. 1.

73.  D. W. Fuerstenau and Pradip, in *Reagents in the Minerals Industry* (M. J. Jones and R. Oblatt, eds.), IMM, London, 1984, p. 161.
74.  D. W. Fuerstenau and S. Raghavan, *Freiberger Forschungsheft, 593*: 75 (1978).
75.  R. W. Smith, *Proc. S. Dak. Acad. Sci., 42*: 60 (1963).
76.  K. Nakatsuka, I. Matsuoka, and J. Shimoiizaka, *Proc. 9th Int. Minerals Processing Cong.*, Prague, 1970, p. 251.
77.  D. W. Fuerstenau, in *Fine Particle Processing*, Vol. 1 (P. Somasundaran, ed.), AIME, New York, 1980, p. 669.
78.  D. W. Fuerstenau and Pradip, in *Reagents in the 3*: 164 (1986).
79.  C. Li, Ph.D. thesis, University of California, Berkeley, 1986.

# V
## SEPARATIONS BASED ON PRECIPITATION

# 12

# Recovery of Surfactant from Surfactant-Based Separations Using a Precipitation Process

LORI L. BRANT,* KEVIN L. STELLNER, and JOHN F.
SCAMEHORN    Institute for Applied Surfactant Research,
University of Oklahoma, Norman Oklahoma

*Present affiliation: Amoco Production Company, Denver, Colorado.
The following organizations are acknowledged for financial sup-
port for this work:  Department of Energy Office of Basic
Energy Sciences Grant no. DE-AS05-84ER13175; the Oklahoma
Mining and Minerals Research Resources Institute; and the
University of Oklahoma Energy Resources Institute.

SYNOPSIS

It is important to recover surfactant from a number of surfactant-based separation processes for reuse in order for the process to be economical.  In this chapter, a new method of recovering ionic surfactant by crystallization or precipitation from a concentrated solution is discussed.  The surfactant can be precipitated using monovalent or multivalent counterions.  In this work, divalent calcium and trivalent aluminum cations were used to precipitate anionic dodecyl sulfate.  Ultimate recoveries of 95% were shown to be attainable using this process.

## I.  INTRODUCTION

When surfactant is added to a system to effect a separation, recovery of the surfactant for reuse may be a requirement for the economic feasibility of a separation process.  The recovered surfactant will normally be recycled to the process to minimize makeup surfactant requirements.

In the removal of dissolved materials from water in many surfactant-based separation processes, a purified water stream and a concentrated aqueous stream result from the process. The concentrated stream may contain both the original dissolved material of interest and the surfactant introduced to the system. Examples of such processes are micellar-enhanced ultrafiltration (Chapters 1 and 2), admicellar chromatography (Chapter 7), and surfactant-enhanced carbon regeneration (Chapter 9).

Recovery of surfactant from this concentrated stream can make a separation process economically attractive by reducing raw material costs.  The material being removed from the original aqueous stream by the surfactant-based separation (e.g., dissolved organic or heavy metal pollutant) is often easier to recover or concentrate further once the surfactant is removed from this concentrated stream.  For example, once surfactant is removed from a retentate system resulting from micellar-enhanced ultrafiltration to remove heavy metals from water (see Chapter 2), the metal cations are no longer bound on micelles and can often be precipitated as hydroxides by pH adjustment.

In this chapter, a recovery process for ionic surfactants is described in general.  Then, specific conditions for recovery of anionic surfactants are given and experimental data shown that describe the effectiveness of this process in recovering sodium dodecyl sulfate (SDS) from an aqueous stream.

## II.  PROCESS DESCRIPTION

### A.  General

The surfactant is precipitated from the aqueous solution by addition of an ion of opposite charge to that of the surfactant (the counterion). The precipitate is removed from solution by gravity settling, filtering, or centrifuging. This filter cake is then recycled back to the process or treated further (as will be described) if the counterion used for the precipitation is unacceptable in the process.

SDS will be used as the model surfactant in subsequent discussions. However, the general principles discussed will apply for other anionic or cationic surfactants.

### B.  Use of a Monovalent Counterion

The surfactant as purchased will normally be in a salt form with a monovalent counterion. Conceptually, the simplest process to precipitate the surfactant from solution is to add this same counterion in high concentrations. The Krafft temperature is increased to a level above the operating temperature or the $K_{SP}$ of the following precipitation reaction is exceeded:

$$C_{12}SO_4^{-} (aq) + Na^{+}(aq) \longrightarrow NaC_{12}SO_4(s) \qquad (1)$$

where $C_{12}SO_4^{-}$ is the dodecyl sulfate anion, $Na^{+}$ is the cationic sodium ion, and $NaC_{12}SO_4$ is the neutral SDS salt. Solid or precipitate is designated by (s) and dissolved species by (aq).

This process is shown schematically in Fig. 1. If the material removed in the surfactant-based separation is an organic or metal that is still soluble in the water after the surfactant has been removed, that material will be emitted from the surfactant precipitation/recovery process with the water. If the material is a liquid organic that is less dense than water and exceeds its solubility, after removal of the surfactant, an organic layer forms on the top of the solution and some fraction of the organic can be directly recovered in high purity.

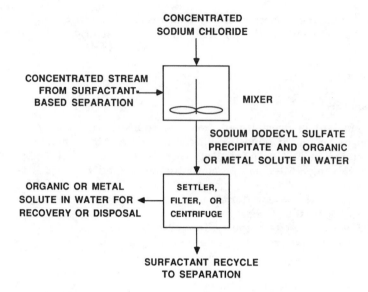

FIG. 1   Process flow diagram for recovery ρf sodium dodecyl sulfate using sodium counterion.


However, if the organic or metal precipitates or forms a liquid that is more dense than water, a clean separation between the precipitated surfactant and the other components is not achieved.

An advantage of this process compared to those to be described using multivalent counterions is that large crystals can form, potentially permitting crystal recovery by the relatively inexpensive process of gravity settling. The main disadvantage of this process is the high concentrations of salt that may be required to precipitate the surfactant. For example, at 30°C, 0.9 M NaCl is required to precipitate the dodecyl sulfate over a wide range of surfactant concentrations (1).

The precipitated surfactant that is recovered will retain some concentrated salt solution in the filter cake. Therefore, some sodium chloride will be added to the process when the surfactant is recycled. The low levels of sodium chloride added back to the process will generally not be a problem.

## C.   Use of a Multivalent Counterion

The solubility product between an ionic surfactant and a multivalent counterion will generally be quite low. As a result, addition of a multivalent counterion in concentrations in slight excess of stoichiometric conditions can potentially cause precipitation of a high proportion of the surfactant. For example, in this work, dodecyl sulfate was precipitated by addition of divalent calcium as soluble calcium chloride:

$$2C_{12}SO_4^- \text{ (aq)} + Ca^{2+}\text{(aq)} \longrightarrow Ca(C_{12}SO_4)_2\text{(s)} \tag{2}$$

The calcium dodecyl sulfate precipitate may be removed from the resulting solution as was the precipitate for the monovalent salt. The same comments concerning removal of the nonsurfactant species during the initial recovery step [reaction (1)] as made for use of a monovalent counterion still apply.

However, this calcium dodecyl sulfate has a very low solubility in water and can't be directly recycled to the surfactant-based separation. It must be transformed to the monovalent salt before it can be redissolved in water for recycle. One way of accomplishing this is to introduce a co-ion (ion of same charge as the surfactant) and water to the multivalent counterion precipitate, in which the co-ion tends to precipitate with the counterion preferentially to the surfactant ion. For example, calcium carbonate is less soluble than the calcium dodecyl sulfate, so addition of an aqueous solution of sodium carbonate to the calcium dodecyl sulfate (after it had been removed from the first step in the process) can result in the following reaction:

$$Ca(C_{12}SO_4)_2\text{(s)} + Na_2CO_3\text{(aq)} \longrightarrow CaCO_3\text{(s)}$$
$$+ 2NaC_{12}SO_4\text{(aq)} \tag{3}$$

A calcium carbonate precipitate and a concentrated solution of dissolved SDS results. After removing the calcium carbonate, the SDS solution can be recycled to the process and the calcium carbonate disposed of, used elsewhere in the plant, or sold. The entire process for SDS and calcium is shown schematically in Fig. 2.

Although a divalent counterion was used for the reactions shown here, counterions of higher valence could be used. For example, aluminum (which is predominately trivalent) could be used as the counterion. For the precipitation and redissolution steps, in analogy to reactions (2) and (3), the following reactions apply:

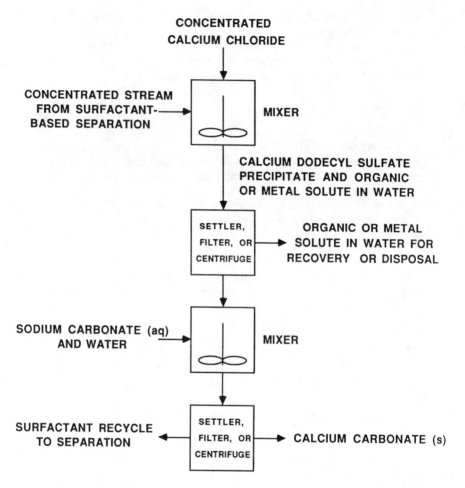

FIG. 2   Process flow diagram for recovery of sodium dodecyl
sulfate using calcium counterion.

$$3C_{12}SO_4^- \text{(aq)} + Al^{3+}\text{(aq)} \longrightarrow Al(C_{12}SO_4)_3\text{(s)} \tag{4}$$

$$2Al(C_{12}SO_4)_3\text{(s)} + 3Na_2CO_3\text{(aq)} \longrightarrow Al_2(CO_3)_3\text{(s)}$$
$$+ 6NaC_{12}SO_4\text{(aq)} \tag{5}$$

The advantage of using a multivalent counterion compared to a monovalent counterion is that a much lower molar quantity is needed to precipitate a specified amount of surfactant. Major disadvantages include the need for steps subsequent to the initial surfactant precipitation step, with associated capital and raw material costs. Also, very fine or colloidal precipitate is almost always observed upon precipitation of surfactants by multivalent counterions, whereas large crystals can be seen when monovalent counterions are used (2). Therefore, multivalent counterions would perhaps preclude the inexpensive gravity settling as a means of removing the surfactant precipitate from the aqueous solution.

Cationic surfactants can be recovered in an analogous fashion to anionic surfactants using either monovalent or multivalent counterions. For example, cationic surfactant cetylpyridinium could be recovered with chloride (monovalent anion) or chromate (divalent anion).

## III. EXPERIMENTAL

The surfactant used in this study, SDS was purified by recrystallization from reagent grade ethanol (plus approximately 50 ml distilled water per 600 ml solution) followed by freeze-vacuum drying. Calcium chloride, aluminum chloride, and sodium carbonate were used as received. SDS, aluminum chloride, and sodium carbonate were obtained from Fisher Scientific Company. Calcium chloride was obtained from Baker Chemical Company. Water used in all experiments was distilled and deionized.

All solutions were made on a weight (molal) basis. Under these conditions, molalities are nearly the same as molarities. Therefore, we report the data in this chapter on a molar basis. A stock solution of SDS was combined with salt solutions of either calcium chloride or aluminum chloride in 15-ml centrifuge tubes to result in 0.2 molal SDS and varying concentrations of $CaCl_2$ (0.04–0.21 molal) and $AlCl_3$ (0.02–0.15 molal). These solutions were shaken until reaction was complete and centrifuged

at 1500 rpm for 20 min. The samples were then decanted and
the SDS concentration in the supernatant was analyzed by
high-performance liquid chromatography (HPLC).

Next, solutions of 0.2 molal SDS and 0.12 molal $CaCl_2$ or
0.08 molal $AlCl_3$ (20% above the stoichiometric ratio of salt to
SDS) were made and allowed to react. The solutions were
centrifuged and the precipitates were then combined with a
known weight of water and $NaCO_3$ to form a slurry to result in
concentrations of 0.03–0.21 molal $Na_2CO_3$ ($CaCl_2$ system) and
0.02–0.14 molal $Na_2CO_3$ ($AlCl_3$ system). The solutions were
well shaken and allowed to react. Analysis of the SDS concentration in the supernatant was again conducted using HPLC.

Experiments were also conducted to determine the rate at
which these reactions occurred. In both the reactions of SDS
with $CaCl_2$ or $AlCl_3$, and the reactions of calcium dodecyl
sulfate or aluminum dodecyl sulfate with $Na_2CO_3$, the supernatant solutions exhibited no change in composition after 5 min.
Approximately 5 min was the least time in which the samples
could be reacted and sufficiently centrifuged to settle the precipitate for decanting. Visually, the precipitate formation appeared to be nearly instantaneous in the reaction of SDS with
calcium chloride and aluminum chloride, and required slightly
longer in the reactions involving sodium carbonate.


## IV.  RESULTS AND DISCUSSION

### A.  Surfactant Precipitation

Solutions containing a nearly constant concentration of SDS and
variable concentrations of calcium chloride were prepared and
the equilibrium SDS concentration measured after equilibration.
By material balance, the fraction of SDS precipitated can be
calculated. The results are shown in Table 1 and Fig. 3.

It is obvious that almost all of the SDS was precipitated
when even a slightly greater than stoichiometric level of calcium
was added, with >99.8% of the surfactant precipitated when
excess counterion was used as seen in Table 1.

The CMC of SDS is 6.6 mM (1) in the absence of added
electrolyte. The concentration of added calcium remaining
after precipitation will cause a depression of the CMC below
this level. However, the CMC will still be well above the 0.3
mM residual concentration observed in Table 1 and Fig. 3 when
more than a stoichiometric amount of calcium is added to

TABLE 1    Fractional Surfactant Precipitation of SDS by Calcium Chloride (Initial [SDS] = 0.197 M)

| Added [calcium chloride] (M) | Initial $\frac{2[Ca^{2+}]}{[SDS]}$ | Residual [SDS] (mM) | % SDS precipitated |
|---|---|---|---|
| 0.046 | 0.467 | 126.1 | 36.0 |
| 0.071 | 0.721 | 56.5 | 71.3 |
| Stoichiometric level | 1.0 | | |
| 0.113 | 1.15 | 0.294 | 99.85 |
| 0.122 | 1.24 | 0.234 | 99.88 |
| 0.165 | 1.68 | 0.148 | 99.92 |
| 0.187 | 1.90 | 0.131 | 99.93 |
| 0.212 | 2.15 | 0.114 | 99.94 |

solution (2). Therefore, when an excess amount of calcium is added to the SDS solution, no micelles are present after precipitation. Consequently, the precipitation reaction shown in reaction (2) can be represented by a simple $K_{SP}$ expression:

$$K_{SP} = [C_{12}SO_4^-]^2[Ca^{2+}] \qquad (6)$$

where component activity coefficients have been assumed to be constant over the range studied. The best fit value of $K_{SP}$ from the data points shown is $5.24 \times 10^{-9}$ $M^3$. Using this $K_{SP}$ the predicted values of the residual SDS concentration in solution after precipitation are shown in Fig. 3. This simple concentration-based solubility product expression accounts for the observed results quite well. Refinements could include addition of activity coefficients (2), but such sophistication is unnecessary for the exploratory work being performed here.

A high residual dissolved dodecyl sulfate concentration is present when less than a stoichiometric amount of calcium chloride is added to the solution (Table 1 and Fig. 3). The surfactant concentration is above the CMC. Modeling of this phenomena is complex and will not be attempted. In application

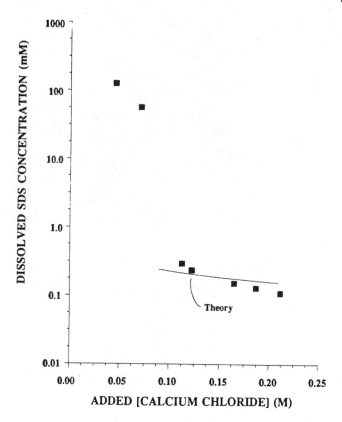

FIG. 3   Residual dissolved SDS concentration after precipitation by calcium.

of this process in industry, an excess of the counterion will
always be added, so that having less than a stoichiometric
amount of counterion is not expected to be of practical interest.

The solubility of an ionic surfactant decreases sharply as
the valence of the counterion with which it is precipitating
increases (2–4).   Therefore, aluminum was also used in the
precipitation step to attempt to exploit this phenomenon as
illustrated by reaction (4).

The precipitation results are shown in Table 2 and Fig. 4.
The equilibrium concentration of the SDS was not as low as in
the calcium case.   Also, the equilibrium dissolved SDS

TABLE 2    Fractional Surfactant Precipitation of SDS by
Aluminum Chloride (Initial [SDS] = 0.196 M)

| Added [aluminum chloride] (M) | Initial $\frac{3[Al^{3+}]}{[SDS]}$ | Residual [SDS] (mM) | % SDS precipitated |
|---|---|---|---|
| 0.022 | 0.337 | 213.1 | −0.5 |
| 0.043 | 0.658 | 44.8 | 77.1 |
| Stoichiometric level | 1.0 | | |
| 0.071 | 1.09 | 0.738 | 99.62 |
| 0.086 | 1.32 | 0.611 | 99.69 |
| 0.107 | 1.64 | 0.571 | 99.71 |
| 0.128 | 1.96 | 0.583 | 99.70 |
| 0.149 | 2.28 | 0.650 | 99.67 |

concentration went through a minima as additional aluminum
chloride was added to solution above the stoichiometric concentration.  This indicates that the apparent solubility of the SDS
was not reduced by adding a trivalent counterion and that the
solubility relationship is not a simple concentration-based $K_{SP}$
equation as in the calcium case.  However, the recovery of the
surfactant using aluminum is still almost quantitative when
excess aluminum is added (>99.6%).

The probable reason for this is that the trivalent aluminum
ion tends to form complexes with chlorides and hydroxides (5).
Variations in the solution pH could result in only a fraction of
the aluminum being present in trivalent form.  A trivalent
cation such as lanthanum, which is trivalent over a wider
range of pH values, might yield more effective results.  In any
case, from a practical viewpoint the precipitation step is more
effectively performed with the divalent calcium than the aluminum.  Additionally, the divalent calcium results in such excellent precipitation characteristics that the incentive to search for
such a trivalent is not high from a practical viewpoint.

## B.  Redissolution of Surfactant

A sodium carbonate solution was added to a slurry of the precipitated calcium dodecyl sulfate in order to allow the dissolution

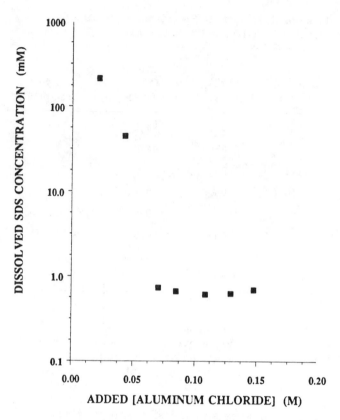

FIG. 4   Residual dissolved SDS concentration after precipitation by aluminum.

reaction (3) to proceed.   The redissolved SDS concentration in solution and percentage recovery of SDS from the original solution is shown in Table 3 and Fig. 5.

With greater than a stoichiometric amount of sodium carbonate added, the recovery of SDS is between 92 and 96% in this step. This indicates the potential for this process to achieve high recovery of surfactant.   Since most of the redissolved SDS is in micellar form, the relationships describing the final equilibrium will involve simultaneous modeling of monomer-micelle equilibrium and the solubility product relationships of the two solids present

TABLE 3    Fractional Redissolution of SDS After Contact of
Calcium Dodecyl Sulfate with Sodium Carbonate (Initial
[Calcium Dodecyl Sulfate] = 0.100 M = 0.200 N)

| Added [sodium carbonate] (M) | Initial $\dfrac{[Na^+]}{[C_{12}SO_4^-]}$ | Residual dissolved [SDS] (M) | % surfactant redissolved |
|---|---|---|---|
| 0.032 | 0.32 | 0.0601 | 30.1 |
| 0.062 | 0.62 | 0.1154 | 57.7 |
| 0.092 | 0.92 | 0.172 | 86.0 |
| Stoichiometric level | 1.0 | | |
| 0.121 | 1.21 | 0.191 | 95.5 |
| 0.150 | 1.50 | 0.185 | 92.5 |
| 0.181 | 1.81 | 0.185 | 92.5 |
| 0.212 | 2.12 | 0.189 | 94.5 |

(calcium dodecyl sulfate and calcium carbonate). This would be
a formidable task and is not attempted here.

A sodium carbonate solution was also added to a slurry of
the precipitated aluminum dodecyl sulfate step in order to
recover the redissolved SDS by reaction (5). The results are
shown in Table 4 and Fig. 5.

When more than a stoichiometric level of the sodium carbonate is added to the aluminum precipitate, the recovery of
SDS is between 89 and 93%. This is comparable to the results
using the divalent calcium instead of trivalent aluminum. However, the advantages speculated to be associated with the high
valence were not observed. The higher the valence of the
counterion, the more colloidal in nature the precipitate formed
and the more difficult it is to remove from solution. Therefore,
one conclusion of this work is that the divalent counterion is
recommended over the trivalent counterion.

Since the precipitation of the surfactant is almost quantitative with either calcium or aluminum counterion, the limit to
ultimate recovery lies in the redissolution step.

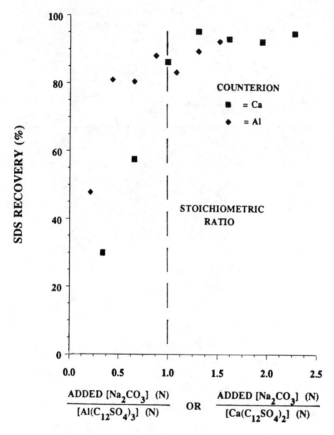

FIG. 5   Percentage recovery of redissolved surfactant after contact with sodium carbonate.

TABLE 4    Fractional Redissolution of SDS After Contact of
Aluminum Dodecyl Sulfate with Sodium Carbonate (Initial
[Aluminum Dodecyl Sulfate] = 0.067 M = 0.200 N)

| Added [sodium carbonate] (M) | Initial $\frac{[Na^+]}{[C_{12}SO_4^-]}$ | Residual dissolved [SDS] (M) | % surfactant redissolved |
|---|---|---|---|
| 0.020 | 0.20 | 0.0956 | 47.8 |
| 0.040 | 0.40 | 0.161 | 80.5 |
| 0.061 | 0.61 | 0.159 | 79.5 |
| 0.081 | 0.81 | 0.177 | 88.5 |
| Stoichiometric level | 1.0 | | |
| 0.102 | 1.02 | 0.166 | 83.0 |
| 0.122 | 1.22 | 0.179 | 89.5 |
| 0.140 | 1.40 | 0.185 | 92.5 |

## V.  CONCLUSIONS AND FUTURE WORK

Approximately 95% recovery of surfactant has been experimentally
observed in preliminary experiments to test the precipitation
scheme for surfactant recovery from surfactant-based separation
processes.  These extremely encouraging results are leading to
future projects to attempt this recovery in the presence of
materials actually removed in the separation process (e.g.,
heavy metals, organic pollutants).  Recovery of these materials
simultaneously with the surfactant will be attempted.  Examina-
tion of the advantages/disadvantages of using a monovalent
counterion [reaction (1)] rather than a divalent counterion
[reaction (2)] for precipitation will be investigated.  Actual
equipment required for the process and the economics of
recovery will be defined.  This ability to recover surfactant
from effluent streams from surfactant-based separations will
often dictate feasibility of these separation processes and
deserves more attention to make this new class of separations
prosper.

REFERENCES

1.  K. L. Stellner and J. F. Scamehorn, *J. Am. Oil Chem. Soc.*, *63*: 566 (1986).
2.  K. L. Stellner and J. F. Scamehorn, Langmuir, in press.
3.  J. M. Peacock and E. Matijevic, *J. Colloid Interf. Sci.*, 77: 548 (1980).
4.  J. Bozic, I. Krznaric, and N. Kallay, *Colloid Polym. Sci.*, *257*: 201 (1979).
5.  R. M. Smith and A. E. Martell, *Critical Stability Constants*, Vols. 4 and 5, Plenum Press, New York, 1976.

# Index